# Fate of Pesticides in Large Animals

EDITED BY

## G. WAYNE IVIE

Veterinary Toxicology and Entomology Research Laboratory
Agricultural Research Service
United States Department of Agriculture
College Station, Texas

## H. WYMAN DOROUGH

Department of Entomology
University of Kentucky
Lexington, Kentucky

ACADEMIC PRESS, Inc.   New York   San Francisco   London   1977
A Subsidiary of Harcourt Brace Jovanovich, Publishers

COPYRIGHT © 1977, BY ACADEMIC PRESS, INC.
ALL RIGHTS RESERVED.
NO PART OF THIS PUBLICATION MAY BE REPRODUCED OR
TRANSMITTED IN ANY FORM OR BY ANY MEANS, ELECTRONIC
OR MECHANICAL, INCLUDING PHOTOCOPY, RECORDING, OR ANY
INFORMATION STORAGE AND RETRIEVAL SYSTEM, WITHOUT
PERMISSION IN WRITING FROM THE PUBLISHER.


ACADEMIC PRESS, INC.
111 Fifth Avenue, New York, New York 10003


*United Kingdom Edition published by*
ACADEMIC PRESS, INC. (LONDON) LTD.
24/28 Oval Road, London NW1


LIBRARY OF CONGRESS CATALOG CARD NUMBER:

ISBN 0-12-376950-7

PRINTED IN THE UNITED STATES OF AMERICA

# Contents

| | |
|---|---|
| Contributors | vii |
| Preface | ix |
| The Toxicological Significance of Large Animal Metabolism Studies with Pesticides | 1 |
|    D. R. FLINT | |
| Techniques and Procedures Used to Study the Metabolism of Pesticides in Large Animals | 17 |
|    G. D. PAULSON | |
| Metabolism Studies of Pesticides in Large Animals: A General Discussion on Regulatory Considerations | 47 |
|    L. C. MISHRA | |
| Comparative Metabolism of Phenoxy Herbicides in Animals | 53 |
|    M. L. LENG | |
| Comparative Metabolism and an Experimental Approach for Study of Liver Oxidase Induction in Primates | 77 |
|    R. I. KRIEGER, J. L. MILLER, S. J. GEE, AND C. R. CLARK | |
| Metabolism of Insect Growth Regulators in Animals | 111 |
|    G. W. IVIE | |
| Comparative Metabolism of Selected Fungicides | 127 |
|    R. C. COUCH AND H. W. DOROUGH | |
| The Kinetics of Halogenated Hydrocarbon Retention and Elimination in Dairy Cattle | 159 |
|    G. F. FRIES | |
| The Metabolism of $p,p'$-DDT and $p,p'$-DDE in the Pig | 175 |
|    G. SUNDSTRÖM, O. HUTZINGER, S. SAFE, AND N. PLATONOW | |

Mirex, Chlordane, Dieldrin, DDT, and PCB's: Metabolites
and Photoisomers in L. Ontario Herring Gulls 183
  
  D. J. HALLETT, R. J. NORSTROM, F. I. ONUSKA,
  AND M. COMBA

DDT Metabolism in Pennsylvania White-Tail Deer 193
  
  D. A. KURTZ AND J. L. GEORGE

The Fate of Phenyl $N,N'$-Dimethylphosphorodiamidate
in Animals 217
  
  R. L. SWANN, M. W. SAUDERHOFF, D. A. LASKOWSKI,
  AND W. H. BRAUN

Metabolism of Croneton (2-Ethylthiomethylphenyl
$N$-Methylcarbamate) in Large Animals 233
  
  H. W. DOROUGH AND D. E. NYE

The Identification of $p$-Nitroaniline as a Metabolite of the
Rodenticide Vacor in Human Liver 253
  
  J. G. OSTERYOUNG, J. W. WHITTAKER, J. TESSARI,
  AND V. BOYES

Index 267

# Contributors

VIRGINIA BOYES, Department of Microbiology, Colorado State University, Fort Collins, Colorado 80523
WERNER H. BRAUN, Dow Chemical USA, Midland, Michigan 48640
CHARLES R. CLARK, Department of Environmental Toxicology, The University of California, Davis, California 95616
MICHAEL COMBA, Canada Centre for Inland Waters, Burlington, Ontario, Canada L7R 4A6
RONALD C. COUCH, Department of Entomology, University of Kentucky, Lexington, Kentucky 40506
H. WYMAN DOROUGH, Department of Entomology, University of Kentucky, Lexington, Kentucky 40506
DONALD R. FLINT, Chemagro Agricultural Division, Mobay Chemical Corporation, Kansas City, Missouri 64120
GEORGE F. FRIES, Pesticide Degradation Laboratory, Agricultural Research Service, U.S. Department of Agriculture, Beltsville, Maryland 20705
SHIRLEY J. GEE, Department of Environmental Toxicology, The University of California, Davis, California 95616
JOHN L. GEORGE, The Pennsylvania State University, University Park, Pennsylvania 16802
DOUGLAS J. HALLETT, Canadian Wildlife Service, Ottawa, Ontario, Canada K1A OH3
OTTO HUTZINGER, Laboratory of Environmental Chemistry, University of Amsterdam, Amsterdam, The Netherlands
G. WAYNE IVIE, Veterinary Toxicology and Entomology Research Laboratory, Agricultural Research Service, U.S. Department of Agriculture, College Station, Texas 77840
ROBERT I. KRIEGER, Department of Environmental Toxicology, The University of California, Davis, California 95616
DAVID A. KURTZ, The Pennsylvania State University, University Park, Pennsylvania 16802
DENNIS A. LASKOWSKI, Dow Chemical USA, Midland, Michigan 48640
MARGUERITE L. LENG, Dow Chemical USA, Midland, Michigan 48640
JEFFREY L. MILLER, Department of Environmental Toxicology, The University of California, Davis, California 95616
LAKSHMI C. MISHRA, U.S. Environmental Protection Agency, Washington, D.C. 20460
ROSS J. NORSTROM, Canadian Wildlife Service, Ottawa, Ontario, Canada K1A OH3
DONALD E. NYE, Thompson-Hayward Chemical Company, Kansas City, Kansas 66110 (present address)
FRANK I. ONUSKA, Canada Centre for Inland Waters, Burlington, Ontario, Canada L7R 4A6
JANET G. OSTERYOUNG, Department of Microbiology, Colorado State University, Fort Collins, Colorado 80523
GAYLORD D. PAULSON, Metabolism and Radiation Research Laboratory, Agricultural Research Service, U.S. Department of Agriculture, Fargo, North Dakota 58102
NICHOLAS S. PLATONOW, Department of Chemistry and Biomedical Sciences, University of Guelph, Guelph, Ontario, Canada N1G 2W1

STEPHEN SAFE, Department of Chemistry and Biomedical Sciences, University of Guelph, Guelph, Ontario, Canada N1G 2W1

MITCHELL W. SAUDERHOFF, Dow Chemical USA, Midland, Michigan 48640

GÖRAN SUNDSTRÖM, Laboratory of Environmental Chemistry, University of Amsterdam, Amsterdam, The Netherlands

ROBERT L. SWANN, Dow Chemical USA, Midland, Michigan 48640

JOHN TESSARI, Department of Microbiology, Colorado State University, Fort Collins, Colorado 80523

JEFFREY W. WHITTAKER, Pfizer Chemical Corporation, Groton, Connecticut 06340 (present address)

# Preface

Studies of the fate of pesticides contribute immensely to the process of evaluating their safety, and to the acquisition of knowledge basic to a greater understanding of their chemistry, biochemistry, and toxicology. The increased emphasis on defining precisely what happens to a pesticide when introduced into the environment, and in various organisms that it may contact therein, is readily apparent by the numerous scientific papers, books, and reports written on the subject in recent years. It may be properly assumed that pesticide fate studies shall be of even greater importance in the future and that their complexity and degree of technical sophistication will attain heights that few would have thought possible just a decade ago.

One cannot help but note the unusually high number of review and state-of-the-art papers dealing with the diverse aspects of pesticides appearing in today's literature. While there may be many reasons for this, one of the most important is the state of flux currently associated with the entire pesticide discipline. Many older compounds are being removed from the marketplace, new restrictions regarding the use of those which remain are constantly being introduced, and products of unique chemical structures and modes of action are appearing on the horizon. It is appropriate then that the pesticide scientist take a close look at where we have been, where we are presently, and more important, where we shall go in the future insofar as chemical control of pests is concerned. It was this general concept and thinking that prompted the organization of a symposium on the fate of pesticides in large animals sponsored by the Pesticide Chemistry Division of the American Chemical Society. This symposium was held at the Centennial Meeting of ACS, August 29–September 3, 1976, in San Francisco, California, and the topics presented at that symposium constitute the substance of this book.

Why a symposium on the fate of pesticides in large animals, and not just in animals generally? The answer to this question is an integral part of several of the papers presented in the symposium but may be summarized by simply stating that large animals (dog, pig, monkey, cow, etc.) often metabolize chemicals differently than small laboratory rodents and these differences (a) may be more representative of the metabolic situation in man, (b) may result in residues in the diet of man not anticipated by small animal studies, and (c) pay provide a basis for the development of concepts pertinent to the phenomenon of selective toxicity. While the differences referred to are a matter of record and for the most part are available to anyone wishing to diligently research the topic, the purpose of the symposium was to focus attention on the fate of pesticides in large animals and to assemble a thesis on the subject covering its theoretical and practical significance.

The symposium was divided into three sections. Papers presented in the first section were designed to delineate the rationale of the symposium and to discuss topics applicable to all facets of large animal metabolism. In the second section, presentations were somewhat more specific in that the comparative metabolism of selected groups of pesticides was considered by the authors. The papers consisted of a general review of the specified subject, followed by the presentation of data from experiments recently conducted by the authors and their colleagues. The third section of the symposium consisted of papers dealing with specific compounds and or specific large animal species. Other than their applicability to the subject of the symposium, no common theme was required, or intended, for those papers presented in the final session. Rather, their purpose was to report results of previously unpublished works representative of the continuing efforts being made to elucidate the fate of pesticides in large animals.

The editors are sincerely appreciative of the efforts and cooperation of the authors, and to the membership of the Pesticide Chemistry Division of the ACS for its support of the symposium. We also extend thanks to our colleagues who assisted in the review and preparation of the manuscripts.

# Fate of Pesticides
in Large Animals

# THE TOXICOLOGICAL SIGNIFICANCE OF LARGE ANIMAL METABOLISM STUDIES WITH PESTICIDES

D. R. Flint

*Chemagro Agricultural Division
Mobay Chemical Corporation*

*ABSTRACT. General agricultural use of pesticides carries with it potential hazards to man directly, by exposure to toxic residues in food, and indirectly, through his environment. Many scientific studies are carried out by pesticide producers to evaluate these potential hazards and thereby establish use conditions consistent with both safety to man and effective pest control for food production. Metabolism studies in large animals, as well as small animals and plants, play an essential role in this hazard evaluation which is shown through the description of interrelationships between these and the overall scientific evaluation program. Some alternatives to large animal metabolism studies and experience with these alternatives are also given.*

INTRODUCTION

The papers in this symposium consider the metabolism of pesticides in large animals. The topic will be covered from many approaches. One might ask why we conduct such studies, what we hope to learn from them, and, of course, to what uses the information will be applied? We must certainly be interested in the nature, extent and rate of biotransformation of the pesticide upon entering the animal organism and whether new or old metabolic pathways were followed. We may be interested in the effect that the route of administration may have on metabolism, absorption, excretion, etc. We would probably be interested in the extent to which the pesticide was absorbed from the gastrointestinal tract following oral administration and whether, upon passing through the intestinal barrier, it continued beyond the liver or was shunted through it *via* the portal and biliary systems back into the intestine. If the pesticide and/or its metabolites passed through the liver into the general circulation, we might ask whether they were bound significantly to plasma proteins or passed into erythrocytes, and to what extent the residues were deposited in tissues bathed by the plasma. Finally, I am sure we would want to know the routes, products, rates and extent of excretion and secretion (e.g., milk). Many more questions may be asked such as, what contribution is made to metabolism by the rumen microflora, or what metabolic differences may be observed under *in vitro* conditions?

Certainly the questions to be answered are many; let us briefly consider some uses of the answers. The nature, extent and rate of metabolism together with the relative toxicity of the metabolites may aid in the design of more effective, yet safer pesticides. Parameters dealing with the dynamics of passage through the organism provide a measure of the likelihood of general or localized residue accumulation, which may aid in the interpretation of toxic effects. More important, the overall scheme of metabolism, excretion and residue accumulation contributes greatly to the evaluation of hazard associated with general use, particularly agricultural use, of the pesticide.

The papers following cover biochemical, physiological and practical aspects of pesticide metabolism in large animals including species comparative metabolism of several classes of pesticides. As these aspects of the subject are amply covered, I will make no attempt to address them as such. Rather, in this initial paper, I shall attempt to apply some measure of general perspective to pesticide metabolism studies in large animals in the hope of bringing together the several specific topics under the general subject of the symposium. I hasten to add that my goal is to provide general perspective to the symposium topic and in no way presume to add to any of the excellent papers to come. To attempt to do so would be not only intolerably pompous, but impossible. This paper is based on an industrial background and as such is not intended to diminish the importance or value of other viewpoints.

TOXICOLOGICAL SIGNIFICANCE AND HAZARD EVALUATION

*General*

The term "toxicological significance" is used in this paper as a means of applying perspective to the subject. It is used in the broad context as relating to the evaluation of human safety. We will consider pesticide metabolism in large animals, in relation to various other specific investigations, as it contributes to the evaluation of human safety associated with pesticide use.

As most of you know, a tremendous amount of research is conducted to demonstrate the safety as well as the efficacy of a pesticide in order to obtain government registration, the official license to manufacture and market a pesticide. Safety applies primarily to man who may be exposed to the toxicant either directly or through his environment. Since most pesticides are at least somewhat toxic to non-target species, including man, a complete evaluation of the potential hazards, particularly toward man, associated with proposed uses of the material is essential. The requirement for demonstration of efficacy also contributes indirectly to the protection of man by limiting the number of potentially hazardous materials available for use.

## Limitation of Pesticide Residues in Foods

Because of their very large and general use, we will concentrate on typical, agricultural-use pesticides although others, not directly applied to food or fiber crops, are also evaluated for safety. The most extensive evaluations, however, are applied to pesticides whose use may result in residues in food or feed.

Pesticides may be applied directly to a growing food crop, reach it by translocation after soil or seed application, or be applied to harvested food crops for protection during storage. Pesticides may also reach human food indirectly by application to forage or fodder which are consumed by animals used for human food. The broadest potential hazard to man, then, is the ingestion of the toxicant or its toxic degradation products with food. It should be noted that although this potential hazard is very general in scope, it is very slight in degree due to the minute amount of toxicant involved. Serious hazard can only be associated with extraordinary exposure to large quantities and/or concentrations of the active agent.

Ideally, the most effective avoidance of this broad potential hazard would be to eliminate all pesticide residues in human food. Unfortunately we have not yet reached that stage of knowledge whereby a toxicant can be designed with toxicity and persistence adequate to control pests during food production, processing and storage, but disappear completely prior to consumption of the food. We can, however, and do minimize the potential for ingestion of toxic residues by controlling the amount and thereby the toxicity of such residues in the human diet. This is accomplished by means of the very many studies mentioned earlier which are performed in connection with application for governmental registration. I would like now to describe these studies and their interrelationships in order to show where and how metabolism studies in large animals contribute to the overall evaluation and minimization of hazard to man associated with pesticide use.

*Direct evaluation of oral toxicity.* One segment of the basic studies used for toxicological evaluation of a pesticide is illustrated in Fig. 1. The most direct means for evaluation of the hazard associated with residues in food involves oral administration of the parent pesticide to animals. Laboratory animals--rats, mice and often dogs--are commonly used in these studies. The parent compound is administered on an acute, sub-chronic or chronic basis with intervals for the latter extending to the full life-span of the animal. Toxicity is measured on the basis of gross, microscopic and biochemical effects observed during and following pesticide administration.

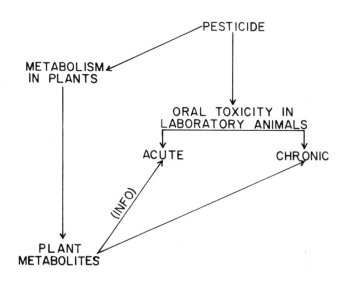

*Fig. 1. Evaluation of oral toxicity of parent pesticide and plant metabolites in laboratory animals.*

*Metabolism studies in the evaluation of oral toxicity.* Fig. 1 shows a second route for oral toxicity evaluation. As the pesticide may be altered by the biochemical systems within

a crop plant producing new chemical species--metabolites--
which may be toxic, it is also necessary to evaluate the
toxicity of the plant metabolites. Specific studies are
conducted with crop plants to determine the nature and extent
of metabolic alteration of the pesticide within the plant
organism. Metabolites identified in these studies which may
be "toxicologically significant" are then administered to
animals on a chronic, or possibly subchronic basis for evaluation of oral toxicity. Acute administration may preceed these
studies for preliminary evaluations of toxicity. The "toxicological significance" judgement is based on information relating chemical structure to toxicity, such as the requirement
for intact organophosphate or carbamate ester functions for
anticholinesterase activity. If sufficient information is not
available for this judgement, the toxicity of the metabolites
must be evaluated experimentally.

It is often found that metabolism results in products
less toxic than the parent compound. It appears to be the
nature of the organism to produce materials of a more polar
nature to facilitate deposition in inert pools in plants, or
elimination by excretion in the case of animals. Reactions
mediating these transformations often result in the destruction of key, functional chemical groups mentioned earlier as
necessary for toxicity. Since it is also known that some
metabolic reactions either do not decrease, or may even
increase the toxicity of a pesticide, it is essential that
metabolic pathways be fully determined in the appropriate crop
plants to permit full toxicological evaluation of pesticide
residues.

Metabolism studies in laboratory animals are also important in the evaluation of oral toxicity. Laboratory animals
are defined here as rats, and dogs in particular since they
are considered in the "Guidelines" as the "usual laboratory
species for extrapolating the metabolism of pesticides to
man."[1] Fig. 2, which is an expansion of Fig. 1, illustrates
the position of laboratory animal metabolism studies in this
evaluation. It becomes apparent in this instance that the
distinction between large and small animals is not as significant as the distinction between laboratory (non-food) animals
and those which form a portion of the human diet.

Metabolism studies are conducted in laboratory animals
to lend support to chronic oral toxicity studies with the
parent pesticide. This support is based on the assumption
that if the animal is capable of producing metabolites from

---

[1]Guidelines for Registering Pesticides in the United
States, U. S. Environmental Protection Agency, *Federal Register*, Vol. 40, Part 162, Sec. 162.81(b) (4)ii(F), June 25, 1975.

the parent compound among which are those produced by the
pertinent plant species, then administration of the parent
compound to animals results in the presence of both parent
pesticide and plant metabolites thereby allowing their
simultaneous and collective toxicological evaluation.

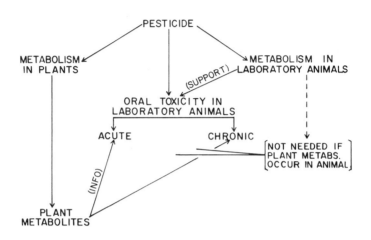

Fig. 2. Evaluation of oral toxicity of parent pesticide and
plant and laboratory animal metabolites in laboratory animals.

In that case the direct evaluation of plant metabolites on a
chronic basis, as shown in Figs. 1 and 2, would be unneces-
sary. If, of course, plant metabolites are not produced in
the animal, separate, chronic toxicity studies may be needed.
Support given to oral toxicity by laboratory animal metabolism
studies is also based on the extrapolation to man mentioned
above.

Metabolism studies in food animals have been added to the
scheme in Fig. 3. These animal species, e.g., chickens, swine,
goats, sheep and cattle, are all representative of the "large
animal" category. Since these animals would normally receive
an agricultural pesticide (or its metabolites) via residues
in the diet, it seems reasonable to carry out metabolism
studies in these species using major plant pesticide residues
in addition to the parent compound. On the other hand, if
the plant metabolites are generated in the animal from the
parent compound, the need to study the animal metabolism of
plant residues is usually considered met.

Frequently metabolic pathways for the pesticide are
identical in animals and plants so that qualitative results

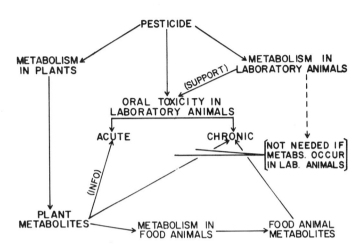

*Fig. 3. Evaluation of oral toxicity in laboratory animals of parent pesticide and plant and animal metabolites.*

are independent of the compound administered. That is, if a plant residue is an intermediary metabolite in animals, its administration to the animal would likely result in its further metabolism along the pathway.

If metabolites found in food animals are qualitatively different from those seen previously in plants and/or laboratory animals, they must also be evaluated for toxicity, as shown in Fig. 3.

The various studies shown in Fig. 3 supply the information needed to evaluate the toxicity of pesticide residues that may be found in foods following agricultural use. Residues produced either in food-crop plants following direct application, or in animals after indirect application (through residue-containing feed) are evaluated for oral toxicity to laboratory animals. This satisfies the first of two approaches to pesticide research mentioned earlier to minimize the likelihood of ingestion of toxic pesticide residues in the human diet. The second approach concerns the amounts or concentrations of toxic pesticide residues that may occur in foods.

*Quantitation of residues and establishment of tolerances.* Metabolism studies conducted in crop plants determine the chemical nature of pesticide residues which may occur in edible crops. An analytical method can then be developed to quantitate these residues in treated crop samples as illustrated in Fig. 4. Crops are then treated in the field according to the use pattern established for pesticidal efficacy and samples harvested for analysis of residue concentrations. The data thus obtained are used to propose the establishment of a residue "tolerance" for that crop. The value of the tolerance is dependent on maximum pesticide residues expected from normal use. A "tolerance" is a legal limitation for specific pesticide residues in specific agricultural commodities set by government in conjunction with registration. Tolerances based on residue data are set for edible portions of the crop as well as for forage if appropriate. Special tolerances called "Food Additive Tolerances" are established for products of food processing activities e.g., corn oil, sugar molasses, cereal bran.

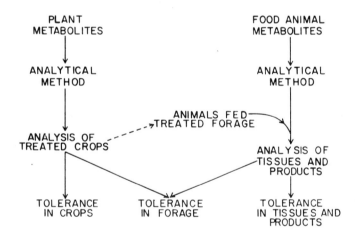

Fig. 4. Steps following metabolism studies taken to obtain residue tolerances.

Using the maximum residue concentration in forage as a guide, food animals are given feed containing pesticide residues on a daily basis for several weeks. The animals are then sacrificed and samples of edible tissues and products (e.g., milk or eggs) are analyzed for residues, the nature of which were determined by the animal metabolism studies. Finally, a tolerance for residues in animal tissues and products is established.

The granting of a tolerance in forage is also dependent

on the feeding study in as much as the residue level in forage may affect the residue levels in animal tissues and products. The latter are limited in magnitude, as will be seen in discussion following, thus posing an indirect limitation to forage tolerances.

The position of the large animal metabolism study is central in importance to obtaining the animal residue tolerance, for it is the metabolism study that yields the nature of the residues that must be quantitated in the feeding test. Quantitation is, of course, dependent on an analytical method which can only be developed when the metabolites have been identified in the appropriate animal species.

*Tolerance limitations - the no-effect level.* Tolerance levels are established not only on the basis of maximum residues found in treated crops or animals. If such were the case they would be limited in magnitude only by the use pattern. Consistent with the assurance of human safety, tolerance levels on commodities directly consumed by humans may not exceed the "No-Effect Level" (NEL) established during toxicity testing. This value is established by feeding various levels of pesticide while monitoring somatic and biochemical effects. The feeding level just below the lowest level producing any real effect is taken as the NEL. Residue tolerances in human foods may not exceed this level. Tolerances in forage are not directly limited by the NEL since direct consumption by man does not apply. They are indirectly limited, however, in that tolerances in animal tissues and products may not exceed the NEL.

A more complex but less direct limitation on tolerance levels exists in the concept of Acceptable Daily Intake (ADI). The ADI is computed as the product of the NEL and standard human body weight divided by a safety factor (10, 100 or 2000 depending on the type of toxicity study used to establish the NEL). The ADI, usually given in micrograms, is the limiting amount of a given pesticide residue which may be present in the human diet on a daily basis. To enforce this limitation another parameter, the Theoretical Daily Intake (TDI) is computed as the sum of the products of individual food tolerances (micrograms/gram) and the statistical amounts[2] of each food (grams) appearing daily in the human diet. The TDI may not exceed the ADI. The TDI is affected by the number and types of foods for which tolerances have been established and as such can limit the number and/or magnitude of additional tolerances which may be granted.

---

[2]A.J.L. (name not given). "The Annual Per Capita Consumption of Selected Items of Food in the United States," *Quarterly Bulletin of the Association of Food and Drug Officials,* Vol. 26, No. 3, July 1962.

*Adjustment of use conditions to limit residues in foods.*
It is by maintaining residue levels below the NEL and within
ADI limitations that conditions are adjusted to minimize
dietary exposure to man. Use conditions can be adjusted to
lower residue levels, if necessary, by lowering application
rates of the pesticide and/or increasing the interval between
the last application and harvest. Obviously, such adjust-
ments must not hamper pest control effectiveness if the
product is to remain viable.

*Toxicological Significance of Pesticide Metabolism Studies in
Animals*

Many studies are thus conducted to establish use conditions for pesticides which will permit the pest control
essential for food production and yet insure safety to humans.
It is clear by the integral positions animal metabolism
studies assume in the overall evaluation program that each has
essential toxicological significance. Without knowledge of
the metabolic pathways taken by the pesticide in laboratory
animals, oral toxicity testing may not be complete as all the
appropriate toxicants may not be tested. Further, if the
metabolic fate of pesticide residues present in the diet of
food animals was not known, it would not be possible either to
evaluate the toxicity or to quantitate the amounts of pesticide residues occurring in animal tissues or products destined
for human consumption. The toxicological significance of
pesticide metabolism studies in both large and small animals
is thus clear for the overall evaluation of safety to humans
associated with the use of agricultural pesticides.
  It is this essential and integral aspect of animal
metabolism studies that I wanted to emphasize. I would like
now to cover some practical aspects of this subject.

PRACTICAL CONSIDERATIONS

*Alternatives to Animal Metabolism Studies*

*Establishment of need.* It seems pertinent to ask whether
the information obtained from such studies is *always* necessary
for hazard evaluation and, if so, can it be acquired by any
other means? Since the need for the information relates to
the entirely pragmatic consideration of potential toxic
residues in human food, the need no longer exists if the
potential for residues does not exist. Thus, those agricultural uses which could not result in pesticide residues in

the diet of food animals would not require a tolerance in animal tissues or products. Not every crop yields forage or by-products which may be used in livestock feed. Thus the need is entirely dependent on residue potential.

The question arises as to the need for a tolerance in animal tissues and products when residues do occur in feed but not in tissues, milk or eggs. In such a case a finite tolerance in the latter is proposed equivalent to the minimum detectable residue i.e., the sensitivity of the analytical method.

Another alternative is available in such cases. If it can be established that no reasonable expectation of residues in animal tissues and products exists under normal use conditions, then a tolerance in the raw agricultural commodity may be established without the necessity of a tolerance in animal commodities. To demonstrate this condition it is usually necessary to furnish proof of the absence of residues in tissues and products following feeding of exaggerated residue levels (10X) in the diet.

*Alternate methods and approaches.* Can the needed information be acquired by any other means? As the primary information needed to support a tolerance is the residue data in tissues and products of fed animals, a metabolism study would not be needed if an analytical method were available for all toxic residues.

Such a method is usually available in the form of radiometric techniques. Radioisotope labeling of the pesticide molecule in appropriate positions can result in labeling of all possible metabolites having toxicological significance. Thus detection of radioactivity in tissue or product samples would be indicative of pesticide residues. Unfortunately this method carries with it the disadvantage of possibly assigning toxicity to residues which may be entirely non-toxic. If such studies result in the presence of radioactive residues, it then becomes necessary to identify the residues in order to evaluate toxicity. Thus, the only advantage to the use of the radiochemical feeding study, in the absence of a corresponding metabolism study, occurs when no residues are detected.

We have made use of this approach on several occasions. One example in which it was particularly useful involved a compound whose structure yielded a substituted aromatic moiety characterized by a *cis* dithiol configuration with the first metabolic reaction. The parent compound was rapidly degraded to this material which in turn had the property of binding tenaciously to biological material. This binding seriously impeded any further investigation of metabolism and the development of analytical methods creating a serious block to registration. The tenacious binding property was exploited however in the assumption that feeding of forage or fodder previously treated with the material would not likely result in tissue or milk residues since no free residues would be available for gastrointestinal absorption. Accordingly, citrus was treated with the radioactive pesticide and, after an appropriate interval on the tree, was harvested and converted to citrus-pulp cattle feed. Continuous feeding of this material to lactating ruminants resulted in no radioactivity in either tissues or milk. Tolerances were established in tissues and milk equivalent to the limit of sensitivity of the radiometric method.

A second example of a general analytical method may be found with a specific type of molecular structure for which all normally possible toxic metabolites can be listed *a priori*. This is illustrated in Fig. 5 by means of a typical organophosphate structure in which one of the ester groups is a substituted aromatic moiety and the other two are small alkyl groups. Plant and laboratory-animal metabolism studies may illustrate that the oxidative reactions shown on phosphorus and sulfur in the figure produce all of the metabolites which would commonly be found containing triesterified phosphorus. The latter is generally required for anticholinesterase activity and is taken as a condition for the major part of the toxicity. If an analytical method can be developed to measure the parent compound plus the five possible metabolites

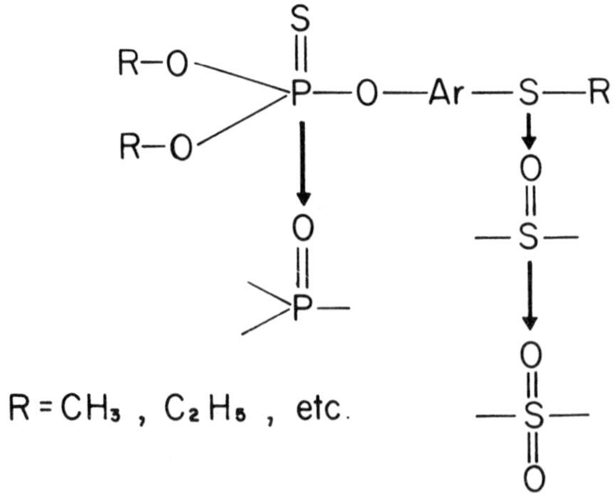

$R = CH_3, C_2H_5$, etc.

Fig. 5. *General structure of an organophosphate pesticide molecule and typical oxidative metabolites.*

resulting from the oxidative reactions illustrated, it may be sufficient for use in a feeding test in the absence of a metabolism study in the corresponding species. Exceptions to this general metabolic scheme may occur, for example, if the aromatic moiety were hydroxylated resulting in toxic metabolites not covered by the method. Experience has shown, and metabolism studies in plants and laboratory animals would be expected to prove, that such metabolites rarely occur in significant concentrations relative to other toxic residues.

We have had experience with an aromatic, substituted phosphoroamidate in which this type of approach would not work. The general structure is shown in Fig. 6. Oxidations on sulfur occurred in several plant and animal systems as well as in soil. No other metabolites above trace concentrations were detected in these systems apart from hydrolysis products. Thus, the development of an analytical method for food animal tissues and products would seem to involve only the three phosphoramidates i.e., sulfide, sulfoxide and sulfone. However, when the metabolism was investigated in the dairy cow, dealkylation on the amide group was discovered requiring a more extensive analytical method in animal tissues and products.

$$R-O \diagdown \atop R-N \diagup P(=O)-O-Ar-S-R$$
$$\phantom{R-}|\phantom{N}$$
$$\phantom{R-N}H$$

$R = CH_3, C_2H_5$, etc.

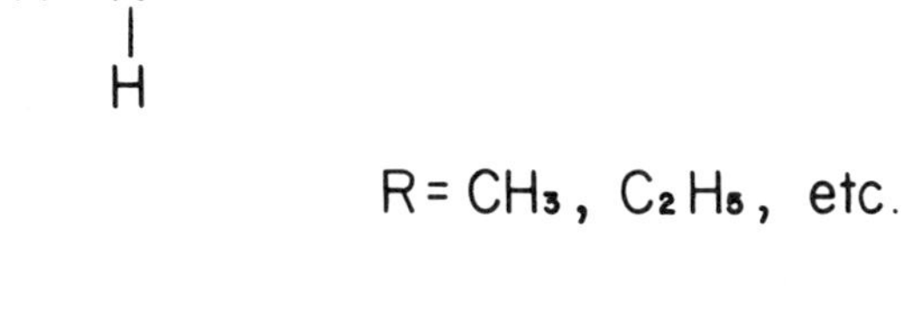

Fig. 6. General structure of an organophosphoramidate pesticide molecule.

*Supplemental Value of Animal Metabolism Data*

Finally, a pesticide producer is continuously subject to inquiries regarding possible intoxications of domestic and wild animals and, occasionally, man. Data reflecting the nature, extent and rate of metabolism, the relative toxicities of the metabolites and the dynamics of passage through the animal organism are extremely helpful in the interpretation of such incidents for the purpose of giving assistance, if needed. Many such incidents are not the result of pesticide exposure, the substantiation of which is also greatly facilitated by metabolism data.

SUMMARY

In summary, the toxicological significance of pesticide metabolism studies in large (and small) animals is proportional to their use in the evaluation of hazards toward man associated with pesticide use. They are integral and essential in the complex program of investigations leading to this evaluation and, therefore, have the highest significance. In some instances the information obtained from certain animal metabolism studies may be obtained by alternate means but this in no way diminishes their toxicological significance.

ACKNOWLEDGMENT

The information presented here is documented to various degrees in the references listed. A great deal of it, however, is based on several years experience in an area which has been termed "Registration Chemistry" and an equivalent amount of interaction with governmental agencies responsible for granting pesticide registrations. This experience is certainly not all mine and in that regard I wish to gratefully acknowledge my most fluent source of information, Dr. T. B. Waggoner, as well as the many others at Chemagro who have contributed their knowledge and experience to this paper.

REFERENCES

Corbett, J. R., "The Biochemical Mode of Action of Pesticides,"
    Academic Press, London and New York, pp 109-164 (1974).
Guidelines for Registering Pesticides in the United States,
    U. S. Environmental Protection Agency, Federal Register,

Vol. 40, Part 162 (1975).
The Regulation of Pesticides in the United States, U. S. Department of Agriculture and U. S. Department of Health, Education and Welfare, Food and Drug Administration (1968).
Tolerances and Exemptions from Tolerances for Pesticide Chemicals in or on Raw Agricultural Commodities, U. S. Environmental Protection Agency, Code of Federal Regulations, Title 40, Chapter I, Subchapter E, Part 180 (1974).

TECHNIQUES AND PROCEDURES USED TO STUDY THE METABOLISM
OF PESTICIDES IN LARGE ANIMALS

G. D. Paulson

Metabolism and Radiation Research Laboratory
Agricultural Research Service
U. S. Department of Agriculture

*ABSTRACT. The experimental equipment, procedures and techniques that may be used to study the metabolism of pesticides in large animals (chickens, sheep, goats, swine, cattle, and horses) are reviewed and discussed. Specific topics that are emphasized include: (1) techniques for administration of pesticides to animals; (2) metabolism stalls and cages for restraining experimental animals; (3) methods and equipment for separate and quantitative collection of excreta, respiratory gases and other volatile products from animals; (4) surgical techniques used to modify animals for metabolic studies (colostomy, cannulations, fistulations, etc.); (5) biopsy techniques; and (6) in vitro techniques that may be useful for studying the metabolism of pesticides in large animals.*

INTRODUCTION

    The rapid increase during the past few decades in the use of pesticides in crop and animal production is well documented. Along with the expanding use of pesticides, there has been a growing concern about the fate of these compounds and their metabolites in animals and animal products consumed by man. A result of this concern has been an expanding interest in the study of the metabolism of pesticides in animals.
    Not surprisingly, most animal metabolism studies with pesticides have been carried out with small animals such as the rat because these animals are cheap, readily available, and easy to work with in the laboratory. However, a review of the literature reveals that the metabolism of a pesticide in the

rat is often different from its metabolism in large animals such as pigs, sheep, and cattle (Paulson, 1975, 1977). Thus, increased emphasis has been on the study of the metabolism of pesticides in meat, milk, and egg-producing animals that may be directly or indirectly exposed to these chemicals.

The purpose of this review is to summarize the techniques, apparatus, and procedures that have been used in large animal pesticide metabolism studies and other procedures and techniques that, in the author's opinion, could be useful in this type of research.

*Animal Management and Facilities*

A complete discussion of laboratory facilities, animal feeding and management, and regulatory requirements for the use of large animals for experimental purposes is beyond the scope of this review. Instead, the reader is referred to excellent and comprehensive reviews on this important topic such as the monograph by Cohen et al. (1972) that includes a discussion of animal management, animal quality and health, and the physical plant required to properly maintain animals, and an excellent bibliography of publications on these and related topics. The recent two-volume series edited by Melby and Altman (1974) includes chapters devoted to housing requirements of large animals, regulatory requirements for handling experimental animals, and other topics on animal experimentation. Descriptions of housing and management techniques for pigs as well as a variety of other topics on the use of pigs in biomedical research have been reviewed in a text edited by Bustad and Burns (1966). The care and management of chickens (Biester, 1953), sheep (Hecker, 1974) and goats (Fletcher et al., 1964) have also been discussed by other authors.

*Dosing Techniques*

The most common procedure for exposing animals to pesticides is via oral treatment. Some workers have mixed radiolabeled pesticides into the ration before feeding the animal (Dorough, 1967; Golab et al., 1969, 1970; Davison, 1970). This method probably best simulates the "field conditions" under which animals are exposed to most pesticides or their plant metabolites. However, accurate quantitation of the administered dose can be a problem, particularly if the pesticide is volatile.

Oral dosing of rats and pigs with nonpolar pesticides by an "ingesta-exchange technique" has been described by Pekas (1974a,b). This procedure involves the incorporation of the chemical into ingesta obtained from a donor animal and the subsequent administration of the mixture by stomach tube to the recipient animals. Pekas (1974a,b) also described a "simulated meal technique" by which the chemical is incorporated into semi-liquid preparations of normal diets and then given to rats and pigs by stomach tube. He reported that these two techniques overcome many of the problems encountered when animals are orally administered a small amount of chemical in solution or in a gelatin capsule. Pekas and Giles (1974) observed that $^{14}$C-carbaryl was absorbed more slowly when incorporated into donor gastric ingesta than when a similar dose was given as an ethanolic solution directly into the stomach of a fasted rat. To the author's knowledge, this technique has been used only with rats and pigs, but there is no apparent reason why it should not be applicable to other animals. The procedure for introducing a stomach tube into horses and ruminants is not difficult (Berge and Westhues, 1966). The technique of force feeding chickens via an oesophageal cannulae (Mather and Ahmad, 1974) would lend itself to the administration of a "simulated meal" that contained the pesticide.

When the pesticide is sufficiently soluble, one convenient and quantitative way to administer it to animals is by stomach tube as an aqueous solution. Some workers have orally administered nonpolar pesticides dissolved in organic solvents. However, the volume of organic solvent given to the animal should be minimized to avoid the possibility of acute or chronic toxicity resulting from the solvent. Some organic solvents may also alter the enzyme system(s) that metabolize pesticides (Feun and Granda, 1970; Conney and Burns, 1972). Other workers have orally dosed animals with pesticides dissolved in polyethylene glycol, olive oil, arachis oil, peanut oil, corn oil, or propylene glycol. Although the use of vehicles such as these has many advantages (for instance, they readily dissolve nonpolar compounds and simplify quantiation), the volume should be kept to a minimum (Pekas and Giles, 1974). One way to avoid the use of solvents is to place the pesticide in a gelatin capsule (without any solvent or place a solution in the capsule and evaporate the solvent) and administer the capsule to large animals such as cows, sheep, and pigs with a "balling gun." Chickens, ducks, and turkeys are also very easily dosed by the capsule technique. For instance, a mature chicken can easily swallow a capsule 1 - 1.5 cm in diameter and 2 - 3 cm in length.

Ruminant animals can be easily and quickly dosed with a pesticide by rumen puncture, a technique that simulates conventional oral dosing and which avoids the mechanical difficulties of dosing them by stomach tube or gelatin capsule and balling gun. With this technique, a large needle is inserted into the rumen about midway between the posterior margin of the rib cage and the hip joint. The pesticide is then injected directly and quantitatively into the rumen (Paulson et al., 1972, 1973). A procedure for injecting materials directly into the abomasum of ruminants has been described by Evans and Spurrell (1967).

When the animal is properly restrained, the test compound can be easily and quickly injected into the jugular vein of horses and ruminants, the ear veins of swine, and the wing or jugular veins of chickens (Berge and Westhues, 1966). However, because most pesticides are nonpolar, dissolving them at the desired concentration in a small volume of a physiological solution is often difficult. When the test compound is dissolved in a nonpolar organic solvent before intravenous injection, acute or chronic toxicity resulting from the organic solvent may be a problem, especially when the volume injected is not kept extremely small. The same problem may be encountered when animals are given intraperitoneal or subcutaneous doses of the test compound dissolved in toxic solvents.

Large animals are exposed to certain pesticides in the gaseous state or as an aerosol. Thus, the metabolism of these pesticides after inhalation may be an important consideration. Loeffler et al. (1976) described apparatus used to expose pigs to an atmosphere of the insecticide dichlorvos. The animals were placed in an airtight stainless-steel cage and air (containing the insecticide) was drawn through the cage. Hutson et al. (1971) described techniques and apparatus used for the treatment of rats by inhalation. They placed the rat in a restraining cage and fitted it with an air-tight headpiece that had one inlet and one exit port. Air was drawn over a microvapor generator and then through the headpiece on the rat. Other researchers studying the effects of cigarette smoke and fractions of this smoke have developed similar methods for inhalation treatment of rabbits (Holland et al., 1958), man (Holland et al., 1959; Egle, 1970), rats (Bretthauer, 1972), and dogs (Egle, 1972). Techniques for the preparation of aerosols and some methods of administration of aerosols to animals (Lippman and Albert, 1968) and man (Morrow, 1974) have been discussed. Modification of these techniques to facilitate inhalation treatment of large farm animals should not be difficult. Also, some of the apparatus designed to administer anesthetics to animals (Markowitz et al., 1964; Berge and Westhues, 1966; Soma, 1971) could likely be used to treat animals with pesticides by inhalation.

Somewhat different procedures for inhalation treatment of rats with radiolabeled insecticides in smoke (Atallah and Dorough, 1974), aerosols (Nye and Dorough, 1976), and solutions of drugs (Burton and Schanker, 1974) have been reported. In brief, with these techniques, the smoke, aerosol, or solution is injected directly into the lungs via the trachea (cannula inserted between the fourth and fifth tracheal ring). Pulmonary absorption and metabolism of radiolabeled insecticides (Atallah and Dorough, 1975; Nye and Dorough, 1976) and drugs (Burton and Schanker, 1974) have been measured in rats by using these techniques. The same concepts and experimental approach should be applicable to the study of pulmonary absorption and metabolism of pesticides in large animals. Techniques for cannulation of the trachea in goats (Fujihara et al., 1973), sheep (Cresswell and Harris, 1961; Young and Webster, 1963), and cattle (Colvin et al., 1957) have been published.

Techniques that were developed for other purposes but which may be useful for pesticide inhalation studies include procedures for the intratracheal inoculation of chickens (Prince et al., 1970) and procedures for introducing inoculations into the nasal passages of horses, sheep, cattle, and pigs (Doll, 1960).

Because animals are exposed to pesticides intentionally (systemic insecticides, etc.) and unintentionally (as a result of drift and contact with forage previously sprayed with herbicides, etc.), there has been some interest in studying the metabolism of these compounds when applied dermally. Methods of dermal application include dipping and spraying (Robbins, et al., 1959; Kaplanis et al., 1959; Krueger et al., 1959; Radeleff and Claborn, 1960), painting (Krueger et al., 1959; Ivie et al., 1976), and the "pour on method." Special precautions must be taken to prevent the animal from licking itself and to avoid "runoff." Accurate quantitation is often difficult because of volatility losses, etc. (Ivie et al., 1976).

*Metabolism Stalls and Other Confinement Units*

Because most pesticide metabolism studies with large animals are conducted with restrained animals, an examination of the stalls and other confinement devices that have been described in the literature is justified.

The adjustable metabolism stall for cattle described by Aschbacher (1972) has been used at this laboratory for pesticide metabolism studies. This unit is adjustable (both length and width) to accommodate different sized animals and is equipped with a removable manger to allow access to the neck and head of the animals. Other metabolism stalls for cattle have been designed and reported by Briggs and Gallup, 1949; Horn et al. (1954), Nelson et al. (1954), Hansard (1951), Seghetti (1953), and Erwin et al. (1956a).

Pekas (1968) described a swine metabolism unit constructed entirely of metal and plastic that can be adjusted to accommodate pigs weighing from 23 to 100 kg. The unit is easily moved, and almost every anatomic region of the pig is accessible to the researcher. Other metabolism units designed for pigs weighing from 20 to 59 kg (Mayo 1961), 39 to 79 kg (Welch et al. 1964), 16 to 79 kg (Beames, 1962), 20 to 79 kg (Morgan and Davey, 1965), and 40 to 138 kg (Hansard et al., 1951) have been described. Metabolism units especially designed for baby pigs (Bell, 1948; Kolari et al., 1955; Sanger et al., 1976), growing pigs (Allen et al., 1963; Braude and Mitchell, 1964; Baker et al., 1967), and lactating sows (Hartman and Pond, 1960) have also been reported.

Metabolism units that are suitable for both sheep and goats have been described by Briggs and Gallup (1949), Hansard (1951), Bratzler (1951), Dick and Mules (1954), Duthie (1959), Till and Downs (1963), and Robbins and Bakke (1967). Metabolism units for horses (Vander Noot et al., 1965; Stillions and Nelson, 1968), and chickens (Richardson et al., 1960; Sanslone and Squibb, 1961; Hart et al., 1969; Paulson, 1969; Guill and Washburn, 1972) have also been reported.

## Separate Collection of Feces and Urine

Although many of the metabolism units described in the previous section were designed to facilitate separate collection of feces and urine, cross contamination of the excreta is often a problem. Thus, a large variety of urine and feces collection devices, usually attached to the animals by means of a harness arrangement, have been developed for cattle (Ballinger and Dunlop, 1946; Hobbs et al., 1950; Balch et al., 1951, 1962; Hansard, 1951; Horn et al., 1954; Putnam et al., 1956; Gorski et al., 1957; Van Es and Vogt, 1959), pigs (Hansard et al., 1951; Kolari et al., 1955; Allen et al., 1963), sheep (Dick and Mules, 1954; Erwin et al., 1959a; Budtz-Olsen et al., 1960; Ingleton, 1971), and horses (Vander Noot et al., 1965). Inflatable urethral catheters have been used to facilitate the collection of urine from female cattle (Cunningham et al., 1955; Aschbacher and Feil, 1968; Aschbacher, 1970), sheep (Aschbacher, 1972; Bakke et al., 1972), goats (Paulson et al., 1973), and pigs (Aschbacher et al., 1971). The catheterization technique is simple to use when the animals are confined to metabolism stalls and, in the author's opinion, is the method of choice when cross contamination of urine and feces is to be completely eliminated. Mellor et al. (1972) reported a technique for catheterization of the bladder of foetal sheep.

There has been limited interest in the study of pesticide metabolism by grazing animals. However, information of this nature may be of value in some instances such as when animals graze pasture treated with herbicides. Apparatus and methods have been reported for the quantitative collection of feces and urine from grazing cattle (Lesperance and Bohman, 1961; Border, 1963) and sheep (Sears and Goodall, 1942; Bassett, 1952; Cook et al., 1952).

Separate collection of urine and feces from the bird is complicated by the mixing of urine and feces in the cloaca. Thus, the bird must be surgically modified or mechanical devices must be fastened to the bird to collect the feces and urine separately. Davis (1927), Coulson and Hughes (1930), and Sturkie and Joiner (1959) collected urine from chickens by using nonsurgical techniques. However, it is the opinion of the author that these methods are not completely satisfactory. For instance, Sturkie and Joiner (1959) observed a decrease in feed and water consumption by birds fitted with a cloacal cannula. Hester et al., (1940), Dixon and Wilkinson (1957), Newberne et al. (1957), Ainsworth (1965), and Elliott and Furneaux (1971) have described surgical procedures for exteriorization of the uretheral openings of the chicken. Several methods for surgically exteriorizing the rectun (colostomy) of the chicken have been reported (Hart and Essex, 1942; Rothchild, 1947; Imabayashi et al., 1956; Fussel, 1960; Richardson et al., 1960; Colvin et al., 1966; Ivy et al., 1968; Tao et al., 1969; Paulson, 1969). Attempts in this laboratory to duplicate some of the earlier colostomy methods were not successful. However, when the modified procedure (Paulson, 1969) was used, 85 to 90 per cent of the surgically prepared chickens were useful for experimental purposes. Feces and urine from birds modified in this way can be separately and quantitatively collected.

## Collection of Respiratory Gases and Other Volatile Products

Quantitative collection of respiratory gases as well as other volatile products such as eructated gases is often necessary in balance studies with radiolabeled pesticides. For instance, the carbon-14 given to animals as some $^{14}$C-carbonyl-labeled carbamate pesticides may be eliminated primarily as $^{14}CO_2$. Volatile products from smaller animals are easily collected by placing the animal in an enclosed metabolism cage and then drawing air through the cage into an appropriate trapping solution(s). This approach has been used for collection of $^{14}CO_2$ from chickens (Paulson, 1969).

Some workers have used respiration masks or "helmets" fitted on the head of large animals such as goats (Hill et al., 1962), sheep (Luick et al., 1960), and cattle (Kleiber and Edick, 1952; Williams, 1960; Yousef and Johnson, 1965). However, if quantitative collection of respiratory gases is required, these devices must be kept on the animal continuously; thus, this approach is satisfactory for only relatively short-term experiments.

Bergman (1964) designed a large respiratory hood for the collection of respiratory gases from sheep. The animal's head was placed in the hood and a collar was placed around the neck of the animal. Air was drawn through the system into collection traps. The animal retained enough mobility so that it could either stand or lie down. Feed and water were placed in the hood; thus, long-term collection of respiratory air was possible. Similar apparatus for long-term collection of respiratory gases from goats (Robbins and Bakke, 1967) and cattle (Rumsey, 1969) has been reported.

One limitation of the techniques described above is that they do not differentiate respiratory gases from gases eructated from the GI tract. Because eructated gases may represent a sizeable proportion of the total volatile products given off by the animals (especially from ruminants), separate collections of the gases from the GI tract and lungs may be important in some metabolism studies. Procedures and techniques for separation of gastrointestinal and respiratory gases in tracheotomized sheep (Cresswell and Harris, 1961; Young and Webster, 1963; Blaxter and Joyce, 1963), goats (Fujihara et al., 1973), and cattle (Colvin et al., 1957; Waldo and Hoernicke, 1961) have been described.

*Collection of Blood*

The collection and assay of circulating blood from animals after administration of a pesticide can give valuable information on absorption, excretion, tissue binding, and kinetics of metabolism. Blood can be easily and quickly collected from the jugular vein of horses and ruminants (Roddick, 1958; Blunt, 1962; Brown, 1963; Berge and Westhues, 1966; Hecker, 1974). When frequent samples are to be collected, and especially when minimal disturbance of the animal is important, the jugular vein may be catheterized by inserting a large bleeding needle into the jugular vein, threading the catheter through the needle into the vein and then removing the needle while holding the catheter in place (Hecker, 1974). A similar procedure has been described by Herd and Barger (1964). The catheter, secured with sutures or tape and irrigated with an anticoagulant to prevent clotting, serves as a convenient means of drawing blood from the test animal. Different types of tubing used for vascular catheters have been discussed by Hecker (1974).

Procedures to exteriorize the carotid artery to facilitate repeated collection of blood from cattle (McDowell et al., 1966; Valli et al., 1967), sheep (Bensadoun and Reid, 1962; Bone et al., 1962; Buck et al., 1965; Hecker, 1974; Klide, 1975), goat (Graham et al., 1937; Buck et al., 1965; Jha et al., 1961), and horses (Tavernor, 1969) have been reported. Catheterization of the tail vein in cattle (Sears and Pape, 1975) has been reported. Procedures have also been described for collection of blood from the cranial vena cava of young pigs (Berge and Westhues, 1966) and adult pigs (Carle and Dewhirst, 1942; Baker and Andreson, 1964; Tegeris et al., 1966).

Heart puncture can be used to obtain large samples of uncontaminated blood from chickens (Andrews, 1944; Genest, 1946; Hofstad, 1950; Krapf, 1959; Garren, 1959; MacArthur, 1960). Other workers have collected blood from the jugular vein of chickens (Law, 1960; Stevens and Ridgway, 1966) and turkeys (Law, 1960), the brachial artery (Philips et al., 1971) and brachial veins (Nordstrom and Smith, 1970) from chickens, and the femoral artery and vein from chickens (Hamre and McHenry, 1942; Rawlings et al., 1970).

In more detailed studies, we may need to collect blood from different sites to obtain information concerning the site of absorption or metabolism. For instance, Katz and Bergman (1969) reported on techniques for cannulation of the hepatic vein, posterior aorta, posterior vena cava, and portal and mesenteric veins of sheep. Thus, they could simultaneously sample blood entering and leaving the liver in the conscious sheep and thereby study the metabolism of compounds by the liver in vivo. This and other techniques for the collection of blood from the pulmonary artery and ventricle of goats (Linzell, 1966), the portal vein of sheep (Jackson et al., 1960; Bensadoun and Reid, 1962; Moodie et al., 1963; Harrison, 1969), and cattle (Conner and Fries, 1960; Carr and Jacobson, 1965; Waldern et al., 1963; Carr and Jacobson, 1968), the hepatic veins of sheep (Moodie et al., 1963; Harrison, 1969), the posterior vena cava of cattle (Hull, 1967), the abdominal aorta of sheep (Yelverton et al., 1969), the adrenal vein in cattle (Lumb et al., 1966), gastrosplenic vein in cattle (Conrad et al., 1958), the mesenteric vein in cattle (Conner and Fries, 1960), the milk vein of goats (Linzell, 1963), the renal vein of sheep (Jackson et al., 1960; Kaufman and Bergman, 1971), and the renal artery of sheep (Jackson et al., 1960) may be useful in some pesticide metabolism studies. A vascular catheter implanting device has been described by Litwak et al., (1975). Additional information on cannulation of the vascular system of ruminants is available in the excellent text by Hecker (1974).

One concern that often arises is whether or not there is a transfer of residues to the fetus in pregnant animals exposed to pesticides. The techniques described by Almond et al., (1970), Comline and Silver (1970), and Soma et al., (1971) make it possible to sample the fetal arterial and venous blood as well as the maternal blood in the conscious sheep. These and related techniques (Gibson and Zemjanis, 1973; Meschia et al., 1969) would be useful to measure the transfer of a pesticide or its metabolites to the fetus.

*Collection of Bile*

Cannulation of the bile duct to facilitate bile collection from the rat has been routinely used for many years by physiologists, biochemists, and others. This technique, when used in conjunction with perfusion studies, has been instrumental in describing the enterohepatic circulation of naturally occurring compounds as well as xenobiotics. Torben (1954) has prepared a monograph on the properties of bile. Although bile has not been so commonly collected from large animals (especially by pesticide chemists), we can reasonably believe that this technique could be effectively used in studies of pesticide metabolism.

Taylor (1960) developed a technique to collect bile uncontaminated with pancreatic juices from sheep. In his procedure, the bile duct was ligated and the gallbladder was cannulated and exteriorized for bile collection. The anterior jejunum was then transected, and the caudal end was cannulated and exteriorized; when bile was not being collected, the two cannulas were connected to restore the flow of bile. Continuity of the intestine was restored by anastomosis anterior to the transection to restore the flow of ingesta through the small intestine. More recently, simpler techniques for collection of bile from sheep have been reported. Gronwall and Cornelius (1970) and Alpert et al. (1969) have described the installation of a "T-tube" (with an inflatable cuff on the distal sidearm) into the common bile duct of sheep. The free end of the "T-tube" was exteriorized and secured to the skin. Bile flowed normally into the duodenum between experiments; however, bile could be quantitatively collected by opening the free end of the "T-tube" and inflating the cuff distal to the liver in order to block any bile flow into the duodenum. Other workers have described the cannulation of the cystic duct (Stanbridge and Mortimer, 1968), hepatic duct (Miyazaki et al., 1972), and common bile duct (Bird, 1972) of the sheep. When bile was not being collected, it was returned to the duodenum via a catheter in the common bile duct in the three latter studies. Boling et al. (1970) cannulated the bile duct below the juncture of the common bile duct and the pancreatic duct and placed a return tube into the duodenum of the sheep. The tubes were brought outside the body and connected, and thus restored the flow of bile and pancreatic secretions into the small intestine when these fluids were not being collected. Leung et al. (1975) placed catheters in the gallbladder and duodenum of a bull to study the enterohepatic circulation of estrone and its metabolites. Hecker (1974) has reviewed several of the procedures used to collect bile from ruminants. Other workers have described related procedures for the collection of bile from pigs (Mahony, 1968) and horses (Gronwall et al., 1975).

The chicken differs from the animals previously discussed in that it has two bile ducts that enter the duodenum separately but which are connected within the liver substance (Clarkson and Richards, 1971). Work by Lind et al. (1967) demonstrated that all of the bile can be excreted by either of the two ducts. Thus, one duct can be ligated and total bile collected by cannulation of the other duct (Clarkson et al., 1957; Lind et al., 1967). At this laboratory, both ducts have been cannulated and exteriorized for bile collection. If one cannula becomes blocked, total bile collection is still possible through the remaining cannula. Similar techniques have been used with turkeys and ducks.

Thus, the methodology for collection of bile from all of the large farm animals has been demonstrated. However, care should be exercized when using these techniques. Dunphy and Stevens (1962) have observed some abnormalities in biliary and hepatic function in animals with grafts in the common bile duct.

*Biopsy Techniques*

Biopsy techniques have not been extensively used in pesticide metabolism studies. However, the use of this technique should be considered, especially when serial samples are of value and when the costs of large animals are important factors. For instance, serial biopsy sampling may be the most efficient and accurate way to measure the kinetics of elimination of tissue-bound pesticide residues. The liver biopsy technique has been perfected and used on a wide variety of animals, including cattle, (Garner, 1950; Loosmore and Allcroft, 1951; Whitehair et al., 1952; Udall et al., 1952; Seghetti and Marsh, 1953; Bone, 1954; Erwin et al., 1956b; Hughes, 1961; Garner et al., 1968; Patterson and Allen, 1968; Oxender et al., 1971), pigs (Erwin et al., 1956; Simmons, 1971; Moser et al., 1972), chickens (Roncalli, 1964; From and Garcia, 1967), and sheep (Dick, 1944). Other techniques that could be useful in the study of pesticide metabolism include methods for biopsy of the mammary gland in cows (Marx and Caruolo, 1963; Hibbit, 1964; Oxender et al., 1971), and goats (Marx and Caruolo, 1963), the kidneys in cattle (Garner et al., 1968), the adipose tissue in cattle (Oxender et al., 1971) and pigs (Erwin et al., 1956b), the rumen of sheep (Dozsa et al., 1963), the skeletal muscle from a variety of large farm animals (Wilson et al., 1955; Bray, 1963; Garner et al., 1968, Harris and Bennett, 1970; Moser et al., 1972; Schmidt et al., 1972), the uterus of cows (Dozsa et al., 1960), testicular tissue (McDonald and Hudson, 1960), the bone marrow of sheep (Grunsell, 1951) and goats (Neal et al., 1962), and skin (Schied et al., 1970).

*Surgically Modified Animals*

Historically, surgically modified animals have not been used extensively in pesticide metabolism studies. However, the fact that such animals have been so widely and successfully used in other physiological and biochemical investigations indicates that pesticide chemists may want to consider these techniques. For instance, cannulation of the GI tract to facilitate collection of samples for assay or as a site for perfusion of test materials may give useful information concerning the metabolism and absorption of a pesticide by different parts of the gut.

A detailed discussion of the procedures, equipment, and reagents used to anesthetize large animals before surgery is beyond the scope of this review. Instead, the reader is referred to previous publications on this important topic (Gregory, 1947; Borrie and Mitchell, 1960; Fisher, 1961; Markowitz et al., 1964; Bergman, 1965; Schaffer, 1965; Berge and Westhues, 1966; Perper and Nazarian, 1967; Piermattei and Swan, 1970; Soma, 1971; Krahwinkel and Evans, 1972; Becker et al., 1972; Ritchie and Hardcastle, 1973; Hecker, 1974). The comprehenisve book edited by Soma (1971) is an excellent reference text and contains detailed descriptions of anesthetics and equipment and procedures for the administration of anesthetics to a variety of animals, including cattle, sheep, goats, pigs, horses, and chickens. Because microsomal enzymes are induced by a wide variety of compounds, including anesthetics and pesticides (Remmer et al., 1968), the researcher must be aware that conducting studies of pesticide metabolism in animals under anesthesia or in animals previously exposed to anesthetics may give misleading results. Perhaps electroanesthesia (Sances and Larson, 1975) would be a satisfactory alternative to chemical anesthetics.

Cannulation of the reticulum (Bost, 1958) and rumen in sheep cattle and goats is easily performed (Nichols, 1953a; Dougherty, 1955; Schnautz, 1957; Brown et al., 1960; Willes et al., 1969; Driedger et al., 1970; Hecker, 1974) and a variety of rumen cannulaes have been described (Ash, 1957; Binns and James, 1959; Yarns and Putnam, 1962; Bowen, 1962; Colvin et al., 1965; Stewart, 1967; Hinkson, 1970). Rumen-fistulated animals can be used for years with only minimal care in addition to that given intact animals. Rumen-fistulated animals can be used as a convenient source of rumen fluid for in vitro studies, and test compounds can be introduced directly into the rumen with less disturbance to the animals than is associated with oral dosing techniques. Moreover, the collection of samples from the rumen after dosing is more convenient when the rumen of the animal is cannulated than when the samples must be collected by stomach tube. Canaway et al. (1965) described a device that automatically sampled the rumen fluid in a fistulated sheep.

Nichols (1953b), Komarek et al. (1960), Hydén (1965), and Foss and Black (1972) have described surgical techniques for preparing an isolated rumen pouch (i.e., no communication between the pouch and the rest of the rumen). Thus, it is possible to study the metabolism and absorption of a test compound or its metabolites in the pouch (mini rumen) in vivo. Related procedures for studying the absorption from an isolated part of the rumen in vivo have been described by Aafjes (1966) and Dobson et al. (1970).

Nutritionists and other workers have been interested in studying the fate of compounds after passage through the stomach into the lower GI tract. This is particularly true for ruminant animals where the rumen microorganisms have such profound metabolic capabilities. Thus, techniques for cannulation of the abomasum of sheep (Ash, 1962; Kondos, 1967; Driedger et al., 1970; Hecker, 1974) and cattle (Hayes et al., 1964; Stewart and Nicola, 1964; Alonso et al., 1973), the omasum of sheep (Bouckaert and Oyaert, 1954; Ash, 1962; Willes and Mendel, 1964; Willes et al., 1969; Tadmor and Neumark, 1972; Hecker, 1974) and goats (Best, 1957), the intestine of cattle (Young, 1951; Conner et al., 1957; Morril et al., 1965), sheep (McDonald, 1953; Dougherty, 1955; Brown et al., 1968; Axford et al., 1971; Hecker, 1974), horses (Baker et al., 1969), and pigs (Cunningham et al., 1962), the cecum of sheep (McRae et al., 1973), pigs (Redman et al., 1964), and horses (Alexander and Donald, 1949; Teeter et al., 1968; Lowe et al., 1970) have been developed. When the cannulae is installed, samples can be easily collected through the cannulae after oral dosing and the absorption or metabolism of a compound perfused through the cannulae can be easily studied. Other workers have studied the metabolism and absorption of compounds in chickens with part of the intestine (Littlefield et al., 1972) or gizzard (Burrows, 1936) surgically removed. Sauer and Brisson, (1961), Williams et al. (1966), and Anderson (1969) described surgical techniques for removal of the rumen, reticulum, and omasum of cattle (i.e., in essence, converting a ruminant into a monogastric animal).

Isolated intestinal loops (Thiry-Vella loops) in swine (Pekas, 1977), chickens (Newman and Taylor, 1958), and sheep (Hopcroft and Banks, 1965) have been prepared. Such a preparation allows the investigator to perfuse (via exteriorized inlet and outlet catheters) the test compound and thereby study its metabolism and absorption by the isolated section of intestine. This technique has been used to study the intestinal absorption of a variety of pesticidal carbamates in conscious pigs (Pekas, 1974c). It should be noted that retention of nonpolar compounds in tubing commonly used for studies of this type can be a serious problem (Pekas, 1972).

One of the major problems in studying the metabolism of pesticides by ruminants under field or pasture conditions is obtaining an accurate assessment of the amount of pesticide or plant pesticide metabolites consumed by the grazing animals. Perhaps the best way to obtain an accurate measure of the forage actually consumed is with oesophageal-fistulated animals; this is especially true when the animals are grazing mixed native and range pastures. Methods have been reported for the preparation and maintenance of the oesophageal fistula in ruminant animals (Torell, 1954; Cook et al.,1958; Hamilton et al., 1960; Rusoff and Foote, 1961; McManus et al., 1962; Yarns et al., 1964; Hofmeyer and Voss, 1964; Van Dyne and Torell, 1964; Chapman, 1964; Bott and Goding, 1965; Bishop and Froseth, 1970). When the oesophageal fistula is properly installed, the animal can eat and drink normally, but when the fistula is opened, the forage consumed by the animal is diverted through the fistula into a collection bag. Saliva secreted by resting oesophageal-cannulated animals can also be collected (Yarns et al., 1965).

In some cases, it may be necessary to repeatedly enter the abdominal cavity of experimental animals; however, to do so by conventional surgical techniques puts considerable stress on the animal. Phillips (1968) described a technique for implanting a plastic zipper to facilitate access to the abdominal cavity of sheep and cattle. This technique allows serial sampling and observation of the liver, GI tract, blood vessels, and lymph vessels in the abdominal cavity during extended experiments.

Other surgical techniques that may be useful to the pesticide chemist include techniques for cannulation of the ureters (Lloyd and Buck, 1971), pancreatic ducts (Taylor, 1960; Phaneuf, 1961; Wass, 1965a,b; Pekas, 1965; Dal Borgo et al., 1968; Hulan et al., 1972), lymph ducts (Lascelles and Morris, 1961; Cole et al., 1969; Morris and McIntosh, 1971), salivary ducts (McDougall, 1949; Stewart and Stewart, 1961; Sasaki and Umezu, 1962; Piercy, 1972) and the stomach of pigs (Morgan et al., 1971). Excellent comprehensive texts on experimental surgery in large animals (Markowitz et al., 1964; Hecker, 1974) should be considered by anyone preparing surgically modified animals. A text describing the properties of plastics and their use in surgery is available (Block and Hastings, 1967).

*Artificial Rumen Techniques*

In the study of metabolism of pesticides in mature ruminant animals, one must be aware of the unique metabolic system within the animal, i.e., the microbial population within the rumen. Because of the unique conditions in the rumen of these animals (for example, the strong reducing potential), many compounds, including pesticides, may be extensively metabolized before they are absorbed from the GI tract of the animal. To differentiate microbial metabolism from the animal metabolism *per se*, study of the metabolism of a compound by rumen microorganisms in vitro is often useful. The techniques to obtain, process, and maintain rumen fluid for in vitro studies are well documented by Hueter et al. (1958), Baumgardt et al. (1962), Raun and Burroughs (1962), Alexander et al. (1965), Johnson (1966), Lane et al. (1968), Sayre and Van Soest (1972), and others. Although most of the methodology for in vitro studies with rumen microorganisms was originally developed for use in nutritional investigations, the same techniques are generally applicable to studies of pesticides. The micro artificial rumen designed for use with radiolabeled compounds (Dawson et al., 1964) may be especially useful to the pesticide chemist.

A related procedure that may be useful to pesticide chemists, especially those studying the fate of radioactive pesticide residues in plant materials when given to ruminants, is the so-called Vivar technique (Fina et al., 1958, 1962; Johnson, 1966). The apparatus is basically a jar (either glass or stainless-steel) in which a rumen fermentation is conducted while the unit is suspended in the rumen of a fistulated animal. The fermentation within the apparatus is separated from the rumen contents of the animal by a bacteriological membrane. Thus, it is possible to determine the microbial metabolism in a semi-isolated system, but in a system that is in equilibrium with the rumen of the animal. Another related technique involves placing the substrate in a bag made of nondigestible material and then placing the bag in the rumen of a fistulated animal (Van Keuren and Heinemann, 1962; Lusk et al., 1962; Johnson, 1966). Care must be exercised to insure that the mesh of the bag is sufficiently small to prevent the loss of particulate matter. However, when properly conducted, this type of study apparently closely simulates in vivo digestion. Although this technique has been most commonly used to study cellulose digestion, it should also be applicable to the study of metabolism of plant pesticide residues.

*In Vitro Techniques*

In vitro organ perfusion techniques have been effectively used by physiologists, biochemists, pharmacologists, and others to study the metabolism of compounds in organs isolated from the rest of the animal. The utility of this approach and the technology developed to perfuse a wide variety of organs and

tissues is documented in texts on this subject by Ross (1972)
and by Ritchie and Hardcastle (1973). It is not surprising
that the majority of the perfusion studies have been conducted
on organs from small animals such as the rat. However, the
same general concepts and experimental procedures are applicable to the perfusion of organs from large animals. For instance, a number of workers have used the isolated perfused
rumen from the goat to study the metabolism of natural products (McCarthy et al., 1958; Brown et al., 1960; Spahr et al.,
1965). Techniques for the perfusion of isolated livers from
goats (McCarthy et al., 1958; Connolly et al., 1964), pigs
(Eiseman et al., 1964), and cows (Chapman et al., 1960), and
mammary glands from cows (Petersen and Shaw, 1941) have been
described. Perfusion of isolated livers from sheep has been
used to estimate the quantity of diethylstilbestrol removed
from blood during a single passage through the liver and to
determine the nature of the metabolites formed by the liver
(Huber et al., 1972).

Sullivan et al. (1972) studied the metabolism of the insecticide carbaryl by an "in vitro organ maintenance technique"
in which tissues were cut into 2-mm cubes and incubated with
the insecticide in oxygenated medium. They reported that this
technique qualitatively reflected the metabolism of pesticides
by intact animals.

Pesticide metabolism has been studied in homogenates of
liver from sheep (Wright et al., 1970; Noguchi, 1972; Douch,
1974) and cattle (Gutenmann and Lisk, 1969). These techniques
make it possible to investigate the intermediate reaction
products in the metabolism of pesticides. In addition, tissue
homogenates and cell fractions may also be used for the biosynthesis of suspected metabolites. For instance, the 105,000
x g supernate from chicken liver fortified with [$^{35}$S]-sulfate
and cofactors was used to study the conjugation of carbaryl
metabolites (Paulson et al., 1970). Other *in vitro* procedures
that may be useful to study the metabolism of pesticides include techniques to culture bovine mammary tissue (Richmond
and Hood, 1973), techniques to grow chick embryos in vitro
(Boone, 1963), and the everted sac technique to measure absorption from the gut of the chicken (Feardon and Bird, 1967).

*Autoradiography*

The use of whole body autoradiography to study the metabolism of radiolabeled xenobiotics in rats, mice, and other
small animals has been discussed by Shindo (1972), Chasseaud
et al. (1972), and many others. Although obtaining sections
of the entire body of large animals such as cows is, for practical purposes, an impossibility, autoradiography techniques

may still be useful to study the metabolism of radiolabeled pesticides in large animals. For instance, this technique would be a powerful tool to determine the localization of radioactivity in different parts of the kidney or other organs from large animals dosed with a radiolabeled pesticide.

GENERAL DISCUSSION AND COMMENTS

It is apparent that many of the techniques and procedures used to study the metabolism of pesticides in large animals are basically the same techniques that physiologists, biochemists, pharmacologists, and others use to study the metabolism of natural products in those animals. Much of the variance in the techniques used with laboratory animals and large monogastric animals results from the difference in size and temperament of the animals. That is to say, the equipment and techniques used with the laboratory rat and a 100 kg "macro rat" (if it existed) would probably be almost as diverse as the equipment and techniques used with rats and pigs. However, in addition to the differences in size, important anatomical and physiological differences between laboratory animals and large farm animals should not be ignored. For instance, many aspects of digestion, absorption of nutrients from the GI tract, and intermediary metabolism as well as anatomical features in the ruminant animal are distinctly different from those in the monogastric animal. Although the processes of digestion and metabolism in the chicken and rat are similar in many ways, differences in the excretory and reproductive systems of these animals are very pronounced. Thus, the cow or the chicken, and in many cases the pig, cannot be considered as a "macro rat." The experimental design and, therefore, the experimental equipment and techniques used must take this into account.

To anyone familiar with the pesticide literature, it is apparent that many of the large animal techniques discussed in the previous sections have seldom or never have been used to study pesticide metabolism in large animals. There are probably a number of explanations for this fact. Some of the techniques mentioned are applicable to only very specialized problems and are of limited use in pesticide metabolism studies. Some of the techniques may never have any practical application for the pesticide chemist. However, I believe that many techniques that have not been used or are only occasionally used should be given more consideration. More frequent use of surgically modified animals, perfusion techniques, and in vitro techniques would lead to a more complete understanding of pesticide metabolism in large animals.

## REFERENCES

Aafjes, J. H., *Nature (London)* 212, 531 (1966).
Ainesworth, L., *J. Poultry Sci.* 44, 1561 (1965).
Alexander, F., Donald D. E., *J. Comp. Pathol.* 59, 127 (1949).
Allen, M. M., Barber, R. S., Braude, R., Mitchell, K. G., *Inst. Anim. Technol. J.* 15, 103 (1963).
Almond, C. H., Boulos, B. M., Davis, L. E., Mackenzie, J. W., *J. Surg. Res.* 10, 7 (1970).
Alonso, F. R., Donawick, W. J., Hammel, E. P., *Am. J. Vet. Res.* 34, 447 (1973).
Alpert, S., Mosher, M., Shanske, A., Arias, I. M., *J. Gen. Physiol.* 53, 238 (1969).
Anderson, W. D., *Am. J. Vet. Res.* 30, 1631 (1969).
Andrews, F. N., *Poultry Sci.* 23, 542 (1944).
Aschbacher, P. W., *J. Anim. Sci.* 35, 1031 (1972).
Aschbacher, P. W., Feil, V. J., *J. Dairy Sci.* 51, 762 (1968).
Aschbacher, P. W., *J. Anim. Sci.* 31, 741 (1970).
Aschbacher, P. W., Feil, V. J., Zaylskie, R. G., *J. Anim. Sci.* 33, 638 (1971).
Ash, R. W., *J. Physiol. (London)* 139, 6P (1957).
Ash, R. W., *Anim. Prod.* 4, 309 (1962).
Ash, R. W., *J. Physiol.* 164, 4 (1962).
Ash, R. W., *Proc. Nutr. Soc.* 28, 110 (1969).
Atallah, Y. H., Dorough, H. W., *J. Agric. Food Chem.* 23, 64 (1975).
Axford, R.F.E., Evans, R. A., Offer, N. W., *Res. Vet. Sci.* 12, 128 (1971).
Baker, D. H., Hiott, W. H., Davis, H. W., Jordan, C. E., *Lab. Practice* 16, 1385 (1967).
Baker, J. P., Sutton, H. H., Crawford, B. H., Lieb, S., *J. Anim. Sci.* 29, 916 (1969).
Baker, L. N., Andresen, E., *Am. J. Vet. Res.* 25, 1559 (1964).
Bakke, J. E., Larson, J. D., Price, C. E., *J. Agric. Food Chem.* 20, 602 (1972).
Balch, C. C., Bartlett, S., Johnson, V. W., *J. Agric. Sci.* 41, 98 (1951).
Balch, C. C., Johnson, V. W., Machin, C., *J. Agric. Sci.* 59, 355 (1962).
Ballinger, C. E., Dunlop, A. A., *N. Z. J. Science Technol.* 27A, 509 (1946).
Bassett, E. G., *N. Z. J. Science Technol.* 34A, 76 (1952).
Baumgardt, B. R., Taylor, M. W., Cason, J. L., *J. Dairy Sci.* 45, 62 (1962).
Beames, R. M., *J. Agric. Sci.* 19, 539 (1962).
Becker, R. M., Lord, L., Dobell, A.R.C., *J. Surg. Res.* 13, 215 (1972).
Bell, J. M., *J. Nutr.* 35, 365 (1948).
Bensadoun, A., Reid, J. T., *J. Dairy Sci.* 45, 540 (1962).

Berge, E., Westhues, M., "Veterinary Operative Surgery,"
   Medical Book Company, Copenhagen, Denmark, 1966.
Bergman, E. N., *Am. J. Vet. Res.* 25, 848 (1964).
Bergman, H. C., *J. Pharm. Sci.* 54, 165 (1965).
Biester, H. E., *Anim. Care Panel Proc.* 4, 98 (1953).
Binns, W., James, L. F., *J. Am. Vet. Med. Assoc.* 135, 603 (1959).
Bird, P. R., *Aust. J. Biol. Sci.* 25, 817 (1972).
Bishop, J. P., Froseth, J. A., *Am. J. Vet. Res.* 31, 1505 (1970).
Blaxter, K. L., Joyce, J. P., *Brit. J. Nutr.* 17, 523 (1963).
Bloch, B., Hastings, G. W., "Plastics in Surgery," Charles C. Thomas, Publisher, Springfield, Ill., 1967.
Blunt, M. H., *Aust. Vet. J.* 38, 397 (1962).
Boling, J. A., Mitchell, G. E., Little, C. O., Fields, C. L., Call, J. L., *J. Anim. Sci.* 29, 504 (1970).
Bone, J. F., *North Am. Vet.* 35, 747 (1954).
Bone, J. F., Metcalfe, J., Parer, J. T., *Am. J. Vet. Res.* 23, 1113 (1962).
Boone, M. A., *Poultry Sci.* 42, 916 (1963).
Border, J. R., Harris, L. E., Butcher, J. E., *J. Anim. Sci.* 22, 521 (1963).
Borrie, J., Mitchell, R. M., *Brit. J. Surg.* 47, 435 (1960).
Bost, J., *J. Physiol. (Paris)* 49, 56 (1957).
Bost, J., *J. Physiol. (Paris)* 50, 170 (1958).
Bott, E. A., Goding, J. R., *Aust. Vet. J.* 41, 326 (1965).
Bouckaert, J. H., Oyaert, W., *Nature (London)* 174, 1195 (1954).
Bowen, J. M., *Am. J. Vet. Res.* 23, 685 (1962).
Bowne, G., Jochim, M. M., Luedke, A. J., *Am. J. Vet. Res.* 25, 561 (1963).
Bratzler, J. W., *J. Anim. Sci.* 10, 592 (1951).
Braude, R., Mitchell, K. G., *Inst. Anim. Techn. J.* 15, 71 (1964).
Bray, R. W., *N. Y. Acad. Sci.* 110, 302 (1963).
Briggs, H. M., Gallup, W. D., *J. Anim. Sci.* 8, 479 (1949).
Brown, G. F., Armstrong, D. G., MacRae, J. C., *Brit. Vet. J.* 124, 78 (1968).
Brown, R. E., Davis, C. L., Staubus, J. R., Nelson, W. O., *J. Dairy Sci.* 43, 1788 (1960).
Buck, W. B., Bond, K., Lyon, N. C., *Cornell Vet.* 55, 154 (1965).
Budtz-Olsen, O. E., Dakin, H. C., Morris, R.J.H., *Aust. J. Agric. Res.* 11, 72 (1960).
Burrows, W. H., *Poultry Sci.* 15, 290 (1936).
Burton, J. A., Schanker, L. S., *Xenobiotica* 4, 291 (1974).
Bustad, L. K., Burns, M. P., Ed., "Swine in Biomedical Research," Frayn Printing Co., Seattle, Wash., 1966.
Canaway, R. J., Terry, R. A., Tilley, J.M.A., *Res. Vet. Sci.* 6, 416 (1965).
Carle, B. N., Dewhirst, W. H., *J. Am. Vet. Med. Assoc.* 101, 495 (1942).

Carr, S. B., Jacobson, D. R., *J. Dairy Sci.* 48, 721 (1965).
Carr, S. B., Jacobson, D. R., *J. Dairy Sci.* 51, 721 (1968).
Chapman, H. W., *Aust. Vet. J.* 40, 64 (1964).
Chapman, N. D., Goldsworthy, P. D., Nyhus, L. M., *Surgery* 48, 111 (1960).
Chasseaud, L. F., Hawkins, D. R., Cameron, B. D., Fry, B. J., Saggers, V. H., *Xenobiotica* 2, 269 (1972).
Clarkson, M. J., Richards, T. G., in "Physiology and Biochemistry of the Domestic Fowl," Vol. 2, Bell., D. J., Freeman, B. M., Ed., Academic Press, New York, 1971, pp. 1085-1114.
Clarkson, T. B., King, J. S., Warnock, N. H., *Am. J. Res.* 18, 187 (1957).
Cohen, B. J., (Chairman), Clarkson, T. B., Flynn, R. J., Hightower, D., Lang, C. M., Melby, E. C., Rabstein, M. M., Soave, O. A., Trum, B. F., Usenik, E. A., Committee on revision of "The Guide for Laboratory Animal Facilities and Care: Guide for the Care and Use of Laboratory Animals," U. S. Government Printing Office, Washington, D. C., 1972.
Colvin, H. W., Noland, P. R., Pharr, L. D., *J. Dairy Sci.*, 48, 995 (1965).
Colvin, H. W., Wheat, J. D., Rhode, E. A., Boda, J. M., *J. Dairy Sci.* 40, 492 (1957).
Colvin, L. B., Creger, C. R., Couch, J. R., Ferguson, T. M., Ansari, M.N.A., *Proc. Soc. Exp. Biol. Med.* 123, 415 (1966).
Comline, R. S., Silver, M., *J. Physiol. (London)* 209, 567 (1970).
Conner, G. H., Fries, G. F., *Am. J. Vet. Res.* 21, 1028 (1960).
Conner, H. G., McGilliard, A. D., Huffman, C. F., *J. Anim. Sci.* 16, 692 (1957).
Conney, A. H., Burns, J. J., *Science* 178, 576 (1972).
Connolly, J. D., Head, H. H., Williams, W. F., *J. Dairy Sci.* 47, 386 (1964).
Conrad, H. R., Smith, H. R., Vandersall, J. H., Pounden, W. D., Hibbs, J. W., *J. Dairy Sci.* 41, 1094 (1958).
Cook, C. W., Stoddart, L. A., Harris, L. E., *J. Anim. Sci.* 11, 578 (1952).
Cook, C. W., Thorne, J. L., Blake, J. T., Edlefsen, J. *J. Anim. Sci.* 17, 189 (1958).
Coulson, E. J., Hughes, J. S., *Poultry Sci.* 10, 53 (1930).
Cresswell, E., Harris, L. E., *Vet. Rec.* 73, 343 (1961).
Cunningham, H. M., Frederick, G. L., Brisson, G. J., *J. Dairy Sci.* 38, 997 (1955).
Cunningham, H. M., Friend, D. W., Nicholson, J.W.G., *Can. J. Anim. Sci.* 42, 112 (1962).
Dal Borgo, G., Harrison, P. C., McGinnis, J., *Poultry Sci.* 47, 1818 (1968).
Davis, R. E., *J. Biol. Chem.* 74, 509 (1927).

Davison, K. L., *J. Agric. Food Chem.* 18, 1156 (1970).
Dawson, R.M.C., Ward, P.F.V., Scott, T. W., *Biochem. J.* 90, 9 (1964).
Dick, A. T., *Aust. Vet. J.* 20, 298 (1944).
Dick, A. T., Mules, M. W., *Aust. J. Agric. Res.* 5, 345 (1954).
Dixon, J. M., Wilkinson, W. S., *Am. J. Vet. Res.* 18, 665 (1957).
Dobson, A., Sellers, A. F., Shaw, G. T., *J. Appl. Physiol.* 28, 100 (1970).
Doll, E. R., *Am. J. Vet. Res.* 21, 518 (1960).
Dorough, H. W., *J. Agric. Food Chem.* 15, 261 (1967).
Douch, P.G.C., *Xenobiotica* 4, 457 (1974).
Dougherty, R. W., *Cornell Vet.* 45, 331 (1955).
Dozsa, L., Anderson, G. C., McLaren, G. A., *Am. J. Vet. Res.* 24, 218 (1963).
Dozsa, L., Olson, N. O., Campbell, A., *Am. J. Vet. Res.* 21, 878 (1960).
Driedger, A., Condon, R. J., Nimrick, K. O., Hatfield, E. E., *J. Anim. Sci.* 31, 772 (1970).
Dunphy, J. E., Stephens, F. O., *Ann. Surg.* 155, 906 (1962).
Duthie, I. F., *Lab. Practice* 8, 408 (1959).
Egle, J. L., *J. Pharmacol. Exp. Ther.* 174, 14 (1970).
Egle, J. L., *Arch. Environ. Health* 25, 119 (1972).
Eiseman, B., Moore, T. C., Normell, L., *Surg. Gynec. Obstet.*, 118, 69 (1964).
Elliott, J. A., Furneaux, R. W., *Poultry Sci.* 50, 1235 (1971).
Erwin, E. S., Dyer, I. A., Ensminger, M. E., Moore, W., *J. Anim. Sci.* 15, 435 (1956a).
Erwin, E. S., Dyer, I. A., Meyer, T. O., Scott, K. W., *J. Anim. Sci.* 15, 428 (1956b).
Erwin, E. S., Blair, J. W., Page, H. M., *J. Anim. Sci.* 18, 935 (1959).
Evans, L., Spurrell, F. A., *J. Appl. Physiol.* 22, 1030 (1967).
Feardon, J. R., Bird, F. H., *Poultry Sci.* 46, 1037 (1967).
Feun, G., Granda, V., *Toxicol. Appl. Pharmacol.* 16, 626 (1970).
Fina, L. R., Keith, C. L., Bartley, E. E., Hartman, P. A., Jacobson, N. L., *J. Anim. Sci.* 21, 930 (1962).
Fina, L. R., Teresa, G. W., Bartley, E. E., *J. Anim. Sci.* 17, 667 (1958).
Fletcher, W. S., Rogers, A. L., Donaldson, S. S., *Lab. Anim. Care* 14, 65 (1964).
Foss, D. C., Black, D. L., *J. Anim. Sci.* 34, 999 (1972).
Fromm, D., Garcia, J. D., *Poultry Sci.* 46, 1536 (1967).
Fujihara, T., Furuhashi, T., Tasaki, I., *J. Dairy Sci.* 56, 820 (1973).
Fussell, M. H., *Nature (London)* 185, 332 (1960).
Garner, H. E., Thurmon, J. C., Romack, F. E., *Am. J. Vet. Res.* 29, 2407 (1968).
Garner, R. J., *Vet. Rec.* 62, 729 (1950).

Garren, H. W., *Poultry Sci.* 38, 916 (1959).
Genest, P., *Can. J. Comp. Med.* 10, 23 (1946).
Gibson, C. D., Zemjanis, R., *Am. J. Vet. Res.* 34, 981 (1973).
Golab, T., Herberg, R. J., Day, E. W., Raun, A. P., Holzer, F. J., Probst, G. W., *J. Agric. Food Chem.* 17, 576 (1969).
Golab, T., Herberg, R. J., Gramlich, J. V., Raun, A. P., Probst, G. W., *J. Agric. Food Chem.* 18, 838 (1970).
Gorski, J., Blosser, T. H., Murdock, F. R., Hodgson, A. S., Soni, B. K., Erb, R. E., *J. Anim. Sci.* 16, 100 (1957).
Graham, W. R., Turner, C. W., Gomez, E. T., *Mo. Agric. Exp. Stn. Bull. No.* 260 (1937).
Gregory, R. A., *Vet. Rec.* 59, 377 (1947).
Gronwall, R., Engelking, L. R., Anwer, M. S., Erichsen, D. F., Klentz, R. D., *Am. J. Vet. Res.* 36, 653 (1975).
Gronwall, R., Cornelius, C. E., *Am. J. Digestive Diseases* 15, 37 (1970).
Grunsell, C. S., *Brit. Vet. J.* 107, 16 (1951).
Guill, R. A., Washburn, K. W., *Poultry Sci.* 51, 1047 (1972).
Gutenmann, W. H., Lisk, D. J., *J. Agric. Food Chem.* 17, 1008 (1969).
Hamilton, F. J., McManus, W. R., Larsen, L. H., *Aust. Vet. J.* 36, 111 (1960).
Hamre, C. J., McHenry, J. T., *Poultry Sci.* 21, 30 (1942).
Hansard, S. L., *Nucleonics* 9, 13 (1951a).
Hansard, S. L. Plumlee, M. P., Hobbs, C. S., Comar, C. L., *J. Anim. Sci.* 10, 88 (1951b).
Harris, R. M., Bennett, J. A., *J. Anim. Sci.* 31, 451 (1970).
Harrison, F. A., *J. Physiol. (London),* 200, 28P (1969).
Hart, S. A., Wilson, W. O., Siopes, T. D., McFarland, L. Z., *Poultry Sci.* 48, 1252 (1969).
Hart, W. M., Essex, H. E., *Am. J. Physiol.* 136, 657 (1942).
Hartman, D. A., Pond, W. G., *J. Anim. Sci.* 19, 780 (1960).
Haslewood, G.A.D., "Bile Salts," Methuen and Company, Limited, London, 1967.
Hayes, B. W., Little, C. O., Mitchell, G. E., *J. Anim. Sci.* 23, 764 (1964).
Hecker, J. F., "Experimental Surgery on Small Ruminants," Butterworth, 1974.
Herd, J. A., Barger, A. C., *J. Appl. Physiol.* 19, 791 (1964).
Hester, H. R., Essex, H. E. Mann, F. C., *Am. J. Physiol.* 128, 592 (1940).
Hibbitt, K. G., *Vet. Rec.* 76, 418 (1964).
Hill, H., Decker, P., Hörnicke, H., Höller, H., Gärtner, K., International Atomic Energy Agency, Conference on "Use of Radioisotopes in Animal Biology and the Medical Sciences," Academic Press, New York, N. Y., 1962, pp. 205-216.
Hinkson, R. S., *J. Anim. Sci.* 34, 779 (1970).

Hobbs, C. S., Hansard, S. L., Barrick, E. R., *J. Anim. Sci.* 9, 565 (1950).
Hofmeyr, C.F.B., Voss, H. C., *J. S. Afr. Vet. Med. Associ.* 35, 579 (1964).
Hofstad, M. S., *J. Am. Vet. Med. Assoc.* 116, 353 (1950).
Holland, R. H., McCall, M. S., Lanz, H. C., *Cancer* 19, 1154 (1959).
Holland, R. H., Wilson, R. H., Morris, D., McCall, M. S., Lanz, H., *Cancer* 11, 709 (1958).
Hopcroft, S. C., Banks, A. W., *Exp. Med. Surg.* 23, 203 (1965).
Horn, L. H., Ray, M. L., Neumann, A. L., *J. Anim. Sci.* 13, 20 (1954).
Huber, T. L., Horn, G. W., Beadle, R. E., *J. Anim. Sci.* 34, 786 (1972).
Hueter, F. G., Gibbons, R. J., Shaw, J. C., Doetsch, R. N., *J. Dairy Sci.* 41, 651 (1958).
Hughes, J. P., *Am. J. Vet. Res.* 23, 1111 (1961).
Hulan, H. W., Moreau, G., Bird, F. H., *Poultry Sci.* 51, 531 (1972).
Hull, M. W., *Am. J. Vet. Res.*, 28, 1915 (1967).
Hutson, D. H., Hoadley, E. C., Pickering, B. A., *Xenobiotica* 1, 593 (1971).
Hydén, S., *Cornell Vet.* 55, 419 (1965).
Imabayashi, K. M., Kametaka, H. T., *Tohoku J. Agric. Res.* 6, 99 (1956).
Ingleton, J. W., *J. Br. Grasslands Soc.* 26, 103 (1971).
Ivie, G. W., Wright, J. E., Smalley, H. E., *J. Agric. Food Chem.* 24, 222 (1976).
Ivy, C. A., Bragg, D. B., Stephenson, E. L., *Poultry Sci.* 47, 1771 (1968).
Jackson, J. B., Radeleff, R. D., Buck, W. B., *J. Am. Vet. Med. Associ.* 136, 440 (1960).
Jha, S. K., Lumb, W. V., Johnston, R. F., *Am. J. Vet. Res.* 22, 948 (1961).
Johnson, R. R., *J. Anim. Sci.* 25, 855 (1966).
Kaplanis, J. N., Hopkins, D. E., Treiber, G. H., *J. Agric. Food Chem.* 7, 483 (1959).
Katz, M. L., Bergman, E. N., *Am. J. Vet. Res.* 30, 655 (1969).
Kaufman, C. F., Bergman, E. N., *Am. J. Vet. Res.* 32, 1103 (1971).
Kleiber, M., Edick, M., *J. Anim. Sci.* 11, 61 (1952).
Klide, M., *Am. J. Vet. Res.* 36, 237 (1975).
Kolari, O. E., Rutledge, E. A., Hanson, L. E., *J. Anim. Sci.* 14, 636 (1955).
Komarek, R. J., Leffel, E. C., Brown, W. H., Mason, K. R., *J. Appl. Physiol.* 15, 181 (1960).
Kondos, A. C., *Aust. Vet. J.* 43, 149 (1967).
Krahwinkel, D. J., Evans, A. T., *J. Am. Vet. Med. Assoc.* 161, 1430 (1972).

Krapf, V., *Berliner und Münchener Tierärztliche Wochenschrift* 72, 363 (1959).
Krueger, H. R., Casida, J. E., Niedermeier, R. P., *J. Agric. Food Chem.* 7, 182 (1959).
Lane, G. T., Noller, C. H., Colenbrander, V. G., Cummings, K. R., Harrington, R. B., *J. Dairy Sci.* 51, 114 (1968).
Lascelles, A. K., Morris, B., *Quart. J. Exp. Physiol.* 46, 199 (1961).
Law, G. R., *J. Poultry. Sci.* 39, 1450 (1960).
Lesperance, A. L., Bohman, V. R., *J. Anim. Sci.* 20, 503 (1961).
Leung, B. S., Pearson, J. R., Martin, R. P., *J. Steroid Biochem.* 6, 1477 (1975).
Lind, G. W., Gronwall, R. R., Cornelius, C. E., *Res. Vet. Sci.* 8, 280 (1967).
Linzell, J. L., *Am. J. Vet. Res.* 24, 223 (1963).
Linzell, J. L., *J. Physiol.* London) 186, 79P (1966).
Lippman, M., Albert, R. E., *J. Air Pollution Control Assoc.* 18, 672 (1968).
Littlefield, L. H., Bletner, J. K., Shirley, H. V., Goff, O. E., *Poultry Sci.* 51, 1721 (1972).
Litwak, P., Lumb, W. V., Butterfield, A. B., *Am. J. Vet. Res.* 36, 109 (1975).
Lloyd, W. E., Buck, W. B., *Am. J. Vet. Res.* 32, 817 (1971).
Loeffler, J. E., Potter, J. C., Scordelis, S. L., Hendrickson, H. R., Huston, C. K., Page, A. C., *J. Agric. Food Chem.* 24, 367 (1976).
Loosmore, R. M., Allcroft, R., *Vet. Rec.* 63, 414 (1951).
Lowe, J. E., Hintz, H. F., Schryver, H. F., *Am. J. Vet. Res.* 31, 1109 (1970).
Luick, J. R., Kleiber, M., Kaneko, J. J., *J. Appl. Physiol.* 15, 175 (1960).
Lumb, W. V., Wolff, W. A., Faulkner, L. C., Parsons, L. E., *Am. J. Vet. Res.* 27, 591 (1966).
Lusk, J. W., Browning, C. B., Miles, J. T., *J. Dairy Sci.* 45, 69 (1962).
MacArthur, F. X., *J. Am. Vet. Med. Assoc.* 116, 38 (1960).
MacRae, J. C., Reid, C.S.W., Dellow, D. W., *Res. Vet. Sci.* 14, 78 (1973).
Mahony, T. D., Berlin, J. D., Pekas, J. C., Sullivan, M. F., "Gastrointestinal Radiation Injury," M. F. Sullivan, Ed., Excerpta Medica Foundation, New York, N. Y., 1968, p. 42.
Markowitz, J., Archibald, J., Downie, H. G., "Experimental Surgery," 5th ed., Williams and Wilkins Company, Baltimore, Md., 1964.
Marx, G. D., Caruolo, E. V., *J. Dairy Sci.* 46, 576 (1963).
Mather, F. D., Ahmad, M. M., *Poultry Sci.* 53, 1610 (1974).
Mayo, R. H., *J. Anim. Sci.* 20, 71 (1961).
McCarthy, R. D., Shaw, J. D., McCarthy, J. L., Lakshmanan, S., Holter, J. B., *Proc. Soc. Exp. Biol. Med.* 99, 556 (1958).

McDonald, I. W., *Vet. Rec.* 65, 290 (1953).
McDonald, L. E., Hudson, R. E., *Am. J. Vet. Sci.* 21, 772 (1960).
McDougall, E. I., *Biochem. J.* 43, 99 (1949).
McDowell, R. E., Underwood, P. C., Lehmann, R. H., Barrada, M. S., *J. Dairy Sci.* 49, 78 (1966).
McManus, W. R., Arnold, G. W., Hamilton, F. J., *Aust. Vet. J.* 38, 275 (1962).
Melby, E. C., Altman, N. H., Ed., "Handbook of Laboratory Animal Science," Vol. I and II, CRC Press, Cleveland, Ohio (1974).
Mellor, D. J., Williams, J. T., Matheson, I. C., *Res. Vet. Sci.* 13, 87 (1972).
Meschia, G., Makowski, E. L., Battaglia, F. C., *Yale J. Biol. Med.* 42, 154 (1969).
Miyazaki, T., Peric-Golia, L., Slaunwhite, W. R., Sandberg, A. A., *Endocrinology* 90, 516 (1972).
Moodie, E. W., Walker, A.I.T., Hutton, P. H., *Q. J. Exp. Physiol.* 48, 379 (1963).
Morgan, D. P., Davey, R. J., *J. Anim. Sci.* 24, 13 (1965).
Morgan, W. W., Gengos, D. C., Talbert, V. L., *Am. J. Vet. Res.* 32, 1879 (1971).
Morrill, J. L., Jacobson, N. L., McGilliard, A. D., Hotchkiss, D. K., *J. Nutr.* 85, 429 (1965).
Morris, B., McIntosh, G. H., *Acta. Endocr. Copenh. Suppl.* 158, 145 (1971).
Morrow, P. E., *Ann. Rev.* Res. Dis. 110, 88 (1974).
Moser, B. D., Peo, E. R., Zimmerman, D. R., Cunningham, P. J., *J. Anim. Sci.* 34, 752 (1972).
Neal, P. A., Rithcie, H. E., Pinsent, P.J.N., *Vet. Rec.* 74, 244 (1962).
Nelson, A. B., Tillman, A. D., Gallup, W. D., MacVicar, R., *J. Anim. Sci.* 13, 504 (1954).
Newberne, P. M., Laerdal, O. A., O'Dall, B. L., *Poultry Sci.* 36, 821 (1957).
Newman, H. J., Taylor, M. W., *Am. J. Vet. Res.* 19, 473 (1958).
Nichols, R. E., *Am. J. Vet. Res.* 14, 35 (1953a).
Nichols, R. E., *Am. J. Vet. Res.* 14, 37 (1953b).
Noguchi, T., "Environmental Toxicology of Pesticides, Proceedings of a U.S.-Japanese Seminar," Oiso, Japan, October, 1971, Matsumura, F., Bousch, G. M., Misata, T., Ed., Academic Press, New York, N. Y., 1972, pp. 607-625.
Nordstrom, J. O., Smith, A. H., *Poultry Sci.* 49, 329 (1970).
Nye, D. E., Dorough, H. W., *Bull. Environ. Contam. Toxicol.* 15, 291 (1976).
Oxender, W. D., Askew, E. W., Benson, J. D., Emery, R. S., *J. Dairy Sci.* 54, 286 (1971).
Patterson, D.S.P., Allen, W. M., *Vet. Rec.* 82, 35 (1968).
Paulson, G. D., *Poultry Sci.* 48, 1331 (1969).

Paulson, G. D., *Res. Rev.* 58, 1 (1975).
Paulson, G. D., Chapter 4 in "Fungicides-Action in Biological Systems, Vol. II," Siegel, M. R., Sisler, H. D., Ed., Marcel Dekker, New York, N. Y., 1977. (In press).
Paulson, G. D., Jacobson, A. M., Zaylskie, R. G., Feil, V. J., *J. Agric. Food Chem.* 21, 804 (1973).
Paulson, G. D., Zaylskie, R. G., Zehr, M. V., Portnoy, C. E., Feil, V. J., *J. Agric. Food Chem.* 18, 110 (1970).
Paulson, G. D., Zehr, M. V., Dockter, M. M., Zaylskie, R. G., *J. Agric. Food Chem.* 20, 33 (1972).
Pekas, J. C., *J. Appl. Physiol.* 20, 1082 (1965).
Pekas, J. C., *J. Anim. Sci.* 27, 1303 (1968).
Pekas, J. C., *Toxicol. Appl. Pharmacol.* 21, 586 (1972).
Pekas, J. C., "Proceedings of the Conference on Food Hygiene and Nutrition," Brno, Czechoslovakia, October 15-17, 1974a, p. 79, abstract.
Pekas, J. C., *Food Cosmet. Toxicol.* 12, 351 (1974b).
Pekas, J. C., *Food Cosmet. Toxicol.* 12, 377 (1974c).
Pekas, J. C., *J. Anim. Sci.* (1977), (In press).
Pekas, J. C., Giles, J. L., *Food Cosmet. Toxicol.* 12, 169 (1974).
Perper, R. J., Najarian, J. S., *Surgery* 61, 824 (1967).
Petersen, W. E., Shaw, J. C., *J. Dairy Sci.* 24, 139 (1941).
Phaneuf, L. P., *Cornell Vet.* 51, 47 (1961).
Phillips, R. W., *Am. J. Vet. Res.* 29, 2423 (1968).
Phillips, R. W., Mitchell, A. D., Nockels, C. F., *Poultry Sci.* 50, 279 (1971).
Piercy, D.W.T., *Res. Vet. Sci.* 13, 383 (1972).
Piermattel, D. L., Swan, H., *J. Surg. Res.* 10, 587 (1970).
Prince, R. P., Fredrickson, T. N., Carrozza, J. H., *Poultry Sci.* 49, 1744 (1970).
Putnam, P. A., Warner, R. G., Loosli, J. K., *J. Dairy Sci.* 39, 1610 (1956).
Radeleff, R. D., Claborn, H. V., *J. Agric. Food Chem.* 8, 437 (1960).
Raun, N. S., Burroughs, W., *J. Anim. Sci.* 21, 454 (1962).
Rawlings, C. A., Krahwinkel, D. J., Anstadt, G. L., *Vet. Med. Small Anim. Clin.* 65, 777 (1970).
Redman, D. R., Teague, H. S., Hendrickx, H. K., King, N. B., *J. Anim. Sci.* 23, 1032 (1964).
Remmer, H., Estabrook, R. W., Schenkman, J., Greim, H., "The enzymatic oxidation of toxicants," Proceedings of a conference held at North Carolina State University at Raleigh, Hodgson, E., Ed., 1968, pp. 65-88.
Richardson, C. E., Watts, A. B., Wilkinson, W. S., Dixon, J. M., *Poultry Sci.* 39, 432 (1960).
Richmond, P. A., Hood, L. F., *J. Dairy Sci.* 56, 611 (1973).

Ritchie, H. D., Hardcastle, J. D., "Isolated Organ Perfusion," University Park Press, Baltimore (1973).
Robbins, J. D., Bakke, J. E., *J. Anim. Sci.* 26, 424 (1967).
Robbins, W. E., Hopkins, T. L., Darrow, D. I., Eddy, G. W., *J. Econ. Entomol.* 52, 214 (1959).
Roddick, B. J., *N. Z. Vet. J.* 6, 20 (1958).
Roncalli, R., *Poultry Sci.* 43, 1357 (1964).
Ross, B. D., "Perfusion Techniques in Biochemistry: A Laboratory Manual in the Use of Isolated Perfused Organs in Biochemical Experimentation," Clarendon Press, Oxford, 1972.
Rothchild, I., *Poultry Sci.* 26, 157 (1947).
Rumsey, T. S., *J. Anim. Sci.* 28, 38 (1969).
Rusoff, L. L., Foote, L. E., *J. Dairy Sci.* 44, 1549 (1961).
Sances, A., Larson, S. J., "Electroanesthesia: Biomedical and Biophysical Studies," Academic Press, New York, 1975.
Sanslone, W. R., Squibb, R. L., *Poultry Sci.* 40, 816 (1961).
Sasaki, Y., Umezu, M., Tohoku J. *Agric. Res.* 13, 211 (1962).
Sauer, F., Brisson, G. J., *Am. J. Vet. Res.* 22, 990 (1961).
Sayre, K. D., Van Soest, P. J., *J. Dairy Sci.* 55, 1496 (1972).
Schaffer, A., Chapter 2 in "Methods of Animal Experimentation," Gay, W. I., Ed., Academic Press, New York and London, 1965.
Schied, R. J., Dolnick, E. H., Terrill, C. E., *J. Anim. Sci.* 30, 771 (1970).
Schmidt, G. R., Zuidam, L., Sybesma, W., *J. Anim. Sci.* 34, 25 (1972).
Schnautz, J. O., *Am. J. Vet. Res.* 18, 73 (1957).
Sears, P. D., Goodall, V. C., *N. Z. J. Sci. Technol.* 23A, 301 (1942).
Sears, P. M., Pape, M. J., *J. Anim. Sci.* 41, 378 (1975).
Seghetti, L., *Am. J. Vet. Res.* 14, 28 (1953).
Seghetti, L., Marsh, H., *Am. J. Vet. Res.* 14, 9 (1953).
Shindo, H., *Ann. Saudyo Res. Lab.* 24, 1 (1972).
Simmons, F., *Am. J. Vet. Res.* 32, 491 (1971).
Smeaton, T. C., Cole, G. J., Simpson-Morgan, M. W., Morris, B., *Aust. J. Exp. Biol. Med. Sci.* 47, 565 (1969).
Soma, L. R., Ed., "Textbook of Veterinary Anesthesia," Williams and Wilkins Company, Baltimore, Md., 1971.
Soma, L. R., White, R. J., Kane, P. B., *J. Surg. Res.* 11, 85 (1971).
Songer, J. R., Mathis, R. G., Skartvedt, S. M., *Am. J. Vet. Res.* 37, 329 (1976).
Spahr, S. L., Kesler, E. M., Flipse, R. J., *J. Dairy Sci.* 48, 228 (1965).
Stanbridge, T. A., Mortimer, P. H., *J. Comp. Pathol.* 78, 499 (1968).
Stevens, R.W.C., Ridgway, G. J., *Poultry Sci.* 45, 204 (1966).
Stewart, W. E., *J. Dairy Sci.* 50, 237 (1967).
Stewart, W. E., Nicolai, J. H., J. Dairy Sci. 47, 654 (1964).

Stewart, W. E., Stewart, D. G., *J. Appl. Physiol.* 16, 203 (1961).
Stillions, M. C., Nelson, W. E., *J. Anim. Sci.* 27, 68 (1968).
Sturkie, P. D., Joiner, W. P., *Poultry Sci.* 38, 30 (1959).
Sullivan, L. S., Chin, B. H., Carpenter, C. P., *Toxicol. Appl. Pharmacol.* 22, 161 (1972).
Tadmor, A., Neumark, H., *Aust, Vet. J.* 48, 408 (1972).
Tao, R., Belzile, R. J., Brisson, G. J., *Am. J. Vet. Res.* 30, 1067 (1969).
Tavernor, W. D., *Am. J. Vet. Res.* 30, 1881 (1969).
Taylor, R. B., *Res. Vet. Sci.* 1, 111 (1960).
Teeter, S. M., Nelson, W. E., Stillions, M. C., *J. Anim. Sci.* 27, 394 (1968).
Tegeris, A. S., Earl, F. L., Curtis, J. M., in "Swine in Biomedical Research," McClellon, R. O., Burns, M. P., Ed., Frayne Printing Co., Seattle, Wa., 1966, pp. 575-596.
Till, A. R., Downs, A. M., *Lab. Pract.* 12, 1006 (1963).
Torell, D. T., *J. Anim. Sci.* 13, 878 (1954).
Udall, R. H., Warner, R. G., Smith, S. E., *Cornell Vet.* 42, 25 (1952).
Valli, V.E.O., McSherry, B. J., Archibald, J., *Can. Vet. J.* 8, 209 (1967).
Vander Noot, G. W., Fonnesbeck, P. V., Lydman, R. K., *J. Anim. Sci.* 24, 691 (1965).
Van Dyne, G. M., Torell, D. T., *J. Range Manage.* 17, 7 (1964).
Van Es, A.J.H., Vogt, J. E., *J. Anim. Sci.* 18, 1220 (1959).
Van Keuren, R. W., Heinemann, W. W., *J. Anim. Sci.* 21, 340 (1962).
Waldern, D. E., Frost, O. L., Harsch, J. A., Blosser, T. H., *Am. J. Vet. Res.* 24, 212 (1963).
Waldo, D. R., Hoernicke, H., *J. Dairy Sci.* 44, 1766 (1961).
Wass, W. M., *Am. J. Vet. Res.* 26, 1103 (1965a).
Wass, W. M., *Am. J. Vet. Res.* 26, 1106 (1965b).
Welch, J. G. Cordts, R. H. Vander Noot, G. W., *J. Anim. Sci.* 23, 183 (1964).
Whitehair, C. K., Peterson, D. R., Van Arsdell, W. J., Thomas, O. O., *J. Am. Vet. Med. Assoc.* 121, 285 (1952).
Willes, R. F., Mendel, V. E., *Am. J. Vet. Res.* 25, 1302 (1964).
Willes, R. F., Mendel, V. E., Robblee, A. R., *J. Anim. Sci.* 29, 661 (1969).
Williams, E. I., Williams, D. E., Goetsch, D. D., Frith, P. O., *Am. J. Vet. Res.* 27, 1777 (1966).
Williams, W. F., *J. Dairy Sci.* 43, 806 (1960).
Wilson, G. D., Batterman, W. E., Sorenson, D. K., Kowalczyk, T., Bray, R. W., *J. Anim. Sci.* 14, 398 (1955).
With, T. K., "Biology of Bile Pigments," Arne Frost-Hanssen Publishers, Copenhagen, 1954.
Wright, F. C., Riner, J. C., Palmer, J. S., Schlinke, J. C., *J. Agric. Food Chem.* 18, 845 (1970).

Yarns, D. A., Putnam, P. A., *J. Anim. Sci.* 21, 744 (1962).
Yarns, D. A., Putnam, P. A., Leffel, E. C., *J. Anim. Sci.* 24, 173 (1965).
Yarns, D. A., Whitmore, G. E., Norcross, M. A., Crandall, M. L., *J. Anim. Sci.* 23, 1046 (1964).
Yelverton, J. T., Henderson, E. A., Dougherty, R. W., *Cornell Vet.* 59, 466 (1969).
Young, B. A., Webster, M.E.D., *Aust. J. Agric. Res.* 14, 867 (1963).
Young, F. W., *J. Am. Vet. Med. Assoc.* 118, 98 (1951).
Yousef, M. K., Johnson, H. D. *J. Dairy Sci.* 48, 104 (1965).

METABOLISM STUDIES OF PESTICIDES IN LARGE ANIMALS:
A GENERAL DISCUSSION ON REGULATORY CONSIDERATIONS

L. C. Mishra[1]

U.S. Environmental Protection Agency

*Abstract. The objective of this report is to examine typical metabolism experiments and suggest how they can be improved for the purpose of safety evaluation. An experiment may often be considered creditable from an academic point of view, but it may contribute little towards safety evaluation. Efforts will be made to emphasize that toxicology and metabolism data in laboratory animals ought to be an integral part in designing, conducting, evaluating and reporting metabolism studies in large animals. The chemical analysis portion of metabolism studies is very important as a methodology and as a tool, but taken out of context of toxicology and safety evaluation, the data loses its usefulness. Thus, mere generation of more chemical data in large animals is not enough; it must be generated and presented in toxicological terms. The inherent difficulties in making decisions regarding the total safety of a pesticide will be discussed.*

Although the economic benefits of pesticides, herbicides, and fungicides, in terms of disease control, increased food production and reduced labor costs have been well recognized for a long time, their potential for producing adverse human health effects resulting from low-level chronic exposure has only recently come to the foreground. There is an extensive debate going on in academic and regulatory circles concerning the toxicity of pesticides and their potential for producing adverse metabolic effects in man. Regulatory agencies are finding it extremely difficult to make decisions regarding

---

[1]The views expressed are those of the author and not necessarily those of the U.S. Environmental Protection Agency.

compounds for which there are conflicting data. Key issues
are: (1) how to extrapolate animal data to humans and (2)
how to determine safe levels for chronic exposure. In view
of this new awareness of possible long term side effects, the
toxicological significance of pesticide residues in food and
feed needs to be reassessed.

The objective of this report is to examine metabolism
studies of pesticides in large animals and discuss, from a
regulatory point of view, which data should be generated and
how it should be presented to best allow regulatory agency
scientists to assess the compound in question. I do not
intend to discuss EPA requirements for registration except
to say that metabolism studies are required if the pesticide
use results in residues in or on food or feed or if a requirement for a tolerance or an exemption from tolerance exists.
I also do not intend to suggest any standard protocol for
metabolism studies, because they can be better accomplished
by the scientists conducting the experiment. My intention is
to identify the problem areas and emphasize the critical role
of toxicology in the design of metabolism studies, and to
stimulate the interest of concerned scientists in this area.

The ideas discussed in this paper therefore should not
be construed as the requirements by EPA for registering pesticides but only as suggestions. For the specific requirements,
one should consult with the Registration Division of EPA.

A cursory review of the metabolism studies in large
animals published during the last six years seems to suggest
that relative to toxicological implications of the data,
overall efforts in this area have been less than desired. In
general, the studies were limited to quantitating elimination
of pesticides in milk, urine, and feces following a single
oral dose, and identification of major metabolites. Some
studies were directed towards accumulation and dissipation of
the parent compound in body tissues and only a few were
involved with specific toxicity problems.

Some of the important regulatory considerations regarding
design of the studies are as follows: (1) objective, scope,
and toxicological basis; (2) selection of dose, schedule, and
route of administration; (3) animal species, strain, sex, age,
and weight; (4) selection of analytical procedure; (5) purity
and characterization of the pesticides; (6) housing conditions.

OBJECTIVE, SCOPE, AND TOXICOLOGICAL BASIS

For any research directed towards investigating safety
of a compound, it is essential that one defines as clearly as
possible the objectives, scope, and toxicological basis of
the planned research. Thus, the resulting data will not only

meet the EPA requirements but may provide additional answers to critical questions related to safety evaluation. Thus philosophy ought to be to define the objectives of the study on the basis of all the available toxicological data and then design the experiments to meet those objectives. For example, one objective may be to determine whether the level of a pesticide or its metabolites will progressively increase in milk during the entire lactation period when cows are exposed to the pesticide under common dairy practices. The protocol and scope to meet this objective should result in more meaningful data than if the objective was only to determine pesticide secretion in milk in dairy cows. Similarly if the objective is to determine tissue distribution of pesticides in relation to blood concentration, absorption, and excretion, the data will be more meaningful than if the only objective is to determine tissue levels. Another goal may be to determine, under conditions of actual use, whether the recommended dosage rate of the pesticide will result in tissue levels estimated to be safe for human consumption. This would require a different experimental design than if the objective is only to gather residue levels in tissues.

One of the objectives of metabolism studies following cutaneous exposure may be to determine cutaneous absorption and excretion of the product in urine and feces. The design of the experiment will need to consider factors such as concentration, pH, temperature, duration of the spray, and the surface area exposed. If an animal is to be dipped in the solution, the duration of the dip, temperature, pH, initial and final concentrations of the dip become important factors in the study.

## SELECTION OF DOSE AND ROUTE OF ADMINISTRATION

These factors are critical to any metabolism study and require careful consideration of toxicological data and estimates of actual exposure of animal species under common agricultural practices. The metabolism patterns and rate of excretion are often dose-dependent. The excretion rate of a pesticide observed after a high dose may not represent that which exists under the actual exposure that occurs following the recommended use. The development of a good data base of animal exposure under common agricultural practices is necessary for making estimates.

## SELECTION OF ANIMAL SPECIES

Selection of animal species for metabolism studies

depends upon the intended use of the product. If the major use involves cows, obviously data generated in pigs will not be useful in evaluating safety because of possible species differences in metabolic pathways in these two species.

## ANALYTICAL PROCEDURES

The selection of an analytical procedure for studying metabolism of pesticides is a very difficult topic and would require an entire symposium to adequately discuss it. Therefore, I will make only a few remarks. One needs to define the objectives of the study very carefully and then select the analytical procedure which will meet the requirements. For example, if the objective is to determine total excretion of the pesticide and its product, then the analytical procedure needs to be the one that can account for the entire molecule and not just one portion of the molecule. This is of particular importance when radioisotope techniques are used. In some cases, analytical procedures such as GLC/MS might be more useful than radioisotope techniques. Other considerations are sensitivity and specificity. The acute and chronic toxicity potential of a compound should be considered in selecting the level of sensitivity of the procedure. A compound toxic at the part per trillion level will need more sensitive methods than a compound toxic at part per million levels. It is also important that preliminary studies are carried out to establish specificity of the procedure. Similarly, the efficiency of recovery of the pesticide from biological materials needs to be established covering the entire range of levels that may be found in tissues. The study will require an adequate number of determinations to permit statistical evaluation of the data.

## PURITY OF THE COMPOUND

Purity of the compound and the quantity and nature of contaminants and decomposition products are very important factors in a study. If a compound is heat, light or humidity sensitive, it will require special care for storage. If the compound is administered via feed, frequent analysis of the feed should be carried out even if the compound appears to be stable under normal use conditions.

## HOUSING CONDITIONS

Last but not the least important are the housing conditions

of the animals. Poor housing conditions may interfere with a metabolism study. It is desirable that one specifies the housing conditions used in the study and if possible and practical to mimic the actual use conditions.

## SOME RESEARCH AREAS RELATED TO SAFETY EVALUATION

There are several areas of research concerning metabolism of pesticides in large animals which have attracted very little attention so far. For example, some pesticides have been shown to produce nitrosamines *in vitro* and epoxides *in vitro* and *in vivo*. Since these metabolites are known to be carcinogenic in animals, possible generation of these materials in animal tissues appears to be important for safety evaluation. Efforts in this area might prove useful in the risk assessment for carcinogenicity of pesticides. Another area of research that needs attention is the identification of residues in terms of loosely bound, covalently bound, or those incorporated into natural metabolites via one carbon pool. Bioavailability studies of covalently bound residues *in vivo* and *in vitro* using rumen fluid and blood may also provide useful information for safety evaluation.

Proper and adequate presentation of the data is important. It is necessary that one follows a consistent pattern in presenting data. Since μg/g is the common expression used in most experimental research, it is more suitable than ppm. Charts and tables need to be concise, clear and self-contained. Results should be critically analyzed with emphasis on validity, significance, and toxicological implications.

In summary, our present technology has given us the capability to detect pesticide residues to almost infinitesimal levels. What we need now is a better understanding of the toxicological significance of residues in food and feed in order to design and conduct more meaningful metabolism studies of pesticides in food producing animals.

## ACKNOWLEDGEMENT

The author is grateful to Dr. Lamar B. Dale, Jr., Dr. Orville E. Paynter, and Mr. Joseph Cummings of the U. S. Environmental Progection Agency for reviewing this paper.

# COMPARATIVE METABOLISM OF PHENOXY HERBICIDES IN ANIMALS

M. L. Leng

*Health and Environmental Research*
*Dow Chemical U.S.A.*

ABSTRACT. *Numerous studies on the metabolism of 2,4-D (2,4-dichlorophenoxy acetic acid) and related herbicides have shown that these compounds are absorbed and distributed rapidly in the body, and are excreted, unchanged, relatively quantitatively in the urine within a week after administration. Pharmacokinetic studies with 2,4,5-T (2,4,5-trichlorophenoxyacetic acid) in rats, dogs, and humans, and with 2,4-D, 2,4,5-T, and silvex [2-(2,4,5-trichlorophenoxy)propionic acid] in humans, corroborated these findings and demonstrated that rates of clearance from plasma and elimination in urine depended on dosage level, animal species, and chemical structure. Dosage levels and chemical structure also affected the pattern of residues found in milk, muscle, fat, liver, and kidney of cattle fed high doses of 2,4-D, 2,4,5-T, silvex, and 2-methyl-4-chlorophenoxyacetic acid (MCPA). Phenol metabolites were found primarily in liver and kidney of cattle and sheep fed high doses of 2,4-D and 2,4,5-T, but were not detected in tissues of animals fed silvex, nor in urine of non-ruminants. The concept of a threshold — a dosage level below which a foreign substance is easily detoxified by biological processes in animals — must be taken into account in extrapolating data from studies at high levels to predict what might be found from exposure of animals to low levels of chemicals, including phenoxy herbicides.*

INTRODUCTION

Summaries of available information on the metabolic fate of phenoxy compounds in animals were included in two recent reviews on degradation of herbicides (Loos, 1975; Paulson, 1975). The purpose of the present paper is to compare studies with four phenoxy herbicides in various species of animals, including man, and to demonstrate the significance of dosage levels on the results obtained in different studies.

The most important phenoxy herbicide is 2,4-D (2,4-dichlorophenoxyacetic acid) widely used for control of broadleaf weeds in field crops, pastures and non-cropland. Minor changes in chemical structure alter the selectivity of herbicidal action. As shown in Fig. 1, substitution of a methyl group for a chlorine at position 2 of the ring gives MCPA which is used primarily in flax, small grains, and certain legumes including peas. Addition of a third chlorine at position 5 gives 2,4,5-T, and further addition of an angular methyl group on the side chain gives silvex. Both trichloro compounds are more effective than 2,4-D in the control of brush and resistant weeds in pastures, rangeland, and non-cropland.

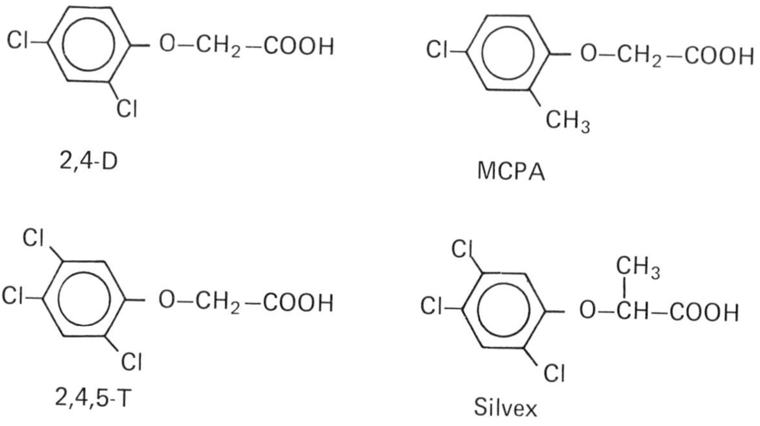

Fig. 1. Chemical structure of four phenoxy herbicides.

The Environmental Protection Agency recently granted additional tolerances for residues of MCPA and 2,4-D in a number of crops and crop byproducts (EPA, 1975, 1976). The tolerances ranged from negligible residues of 0.1 ppm in field crops to 300 or 1000 ppm in or on pasture or range grass at the time of treatment with MCPA or 2,4-D at rates of 2 or 3 lb/A, respectively. Tolerances were also granted for combined residues of these herbicides and their corresponding phenol metabolites in milk and meat. As of this writing, petitions requesting similar tolerances for 2,4,5-T and silvex are still under review in EPA.

This paper includes data on the distribution of residues in milk and edible tissues of cattle maintained on diets containing high levels of these herbicides for 2 to 4 weeks.

## SUMMARY OF STUDIES WITH PHENOXY HERBICIDES IN ANIMALS

Most studies on metabolism of phenoxy herbicides have been done in plants, chiefly to elucidate differences in mode of action of 2,4-D and its analogs and derivatives in various species and varieties. However, numerous investigations have also been conducted in animals after administration of single or repeated doses over a wide range of dosage levels.

Calculations indicate that a human ingesting 0.1 ppm of a chemical in the total daily diet would receive a dose of only 0.0025 milligrams per kilogram of body weight per day. A cow ingesting 1000 ppm in the daily ration would receive a dose of about 30 mg/kg/day.

TABLE 1

*Dosage levels given as single dose (mg/kg of body weight) in studies on metabolism of phenoxy herbicides in animals.*

| Animals | 2,4-D | MCPA | 2,4,5-T | Silvex |
|---|---|---|---|---|
| Mice | 100[1] | 0.8[2] | 100[1] | |
| Rats | 100[3]<br>3 - 300[5]<br>0.04 - 37.5[8]<br>5[4] | | 100[3]<br>40 - 50[6]<br>0.05 - 50[8]<br>50[9]<br>0.17 - 41[10]<br>5 - 200[11]<br>5, 100[12] | 5[4]<br>5, 50[7] |
| Dogs | | | 5[11] | |
| Cattle | 50 - 200[3]<br>0.3[14,15] | 0.3[13] | | |
| Sheep | 4[16] | | 25[17] | |
| Pigs | 50 - 100[3] | | | |
| Chickens | 100, 200[3] | | | |
| Humans | 5[18]<br>2 - 5[20] | | 5[19]<br>2 - 5[20]<br>2[21] | 1[4] |

[1] Zielinski, 1967
[2] Lindquist, 1974
[3] Erne, 1966a
[4] Dow, unpublished
[5] Khanna, 1966
[6] Courtney, 1970
[7] Sauerhoff, 1976c
[8] Shafik, 1971
[9] Grunow, 1971
[10] Fang, 1973
[11] Piper, 1973
[12] Sauerhoff, 1976b
[13] Gutenmann, 1963b
[14] Lisk, 1964
[15] Bache, 1964b
[16] Clark, 1964
[17] Clark, 1971c
[18] Sauerhoff, 1976a
[19] Gehring, 1973
[20] Kohli, 1974
[21] Matsumura, 1970

*Single Doses*

Table 1 lists studies in which single doses of 2,4-D, MCPA, 2,4,5-T, or silvex were administered to mice, rats, dogs, cattle, sheep, pigs, chickens, and humans.

The dosage levels listed in the table are all expressed in terms of milligrams of phenoxy acid per kilogram of body weight for the various species. In some cases the levels shown were calculated from doses expressed in other terms, such as 1 to 80 mg per rat (Khanna and Fang, 1966), or as 5 ppm in 50 lb of feed for cattle (Gutenmann et al., 1963a; Lisk et al., 1963; Bache et al., 1964a). The dosage levels ranged from 0.04 to 300 mg/kg in small laboratory animals, and from 0.3 to 200 mg/kg in domestic animals.

A total of six studies have been conducted in humans at dosage levels ranging from 1 mg/kg for silvex to 5 mg/kg for 2,4-D and 2,4,5-T.

Most doses were given by the oral route, such as in a capsule or by direct intubation into the stomach of experimental animals other than humans. In some cases, the phenoxy herbicides were injected, either subcutaneously in a penetrating solvent such as DMSO (Zielinski and Fishbein, 1967; Courtney, 1970), or intravenously as salt solutions of the phenoxy compounds (Lindquist, 1974; Sauerhoff et al., 1976a,c; Dow, unpublished data). Intraruminal administration has also been used in studies on 2,4-D and 2,4-DB, its $\omega$-substituted butyric acid analog (Gutenmann et al., 1963a,b).

*Repeated Doses*

Numerous studies have also been conducted in animals given daily doses of these herbicides for intervals ranging from 4 days to 2 years. As outlined in Table 2, the dosage levels in rats ranged from 50 mg/kg/day for 5 days (Courtney, 1970) to an estimated 120 mg/kg/day for up to 2 years in animals given 1000 ppm 2,4-D in their drinking water (Erne, 1966a). Known dosage levels in livestock ranged up to 60 mg/kg/day for calves and sheep given 2000 ppm 2,4-D, 2,4,5-T, or silvex in the total diet for 28 days (Leng, 1972). Estimated dosage levels were about 20 mg/kg/day in pigs maintained on a diet containing 500 ppm 2,4-D (Erne, 1966a), and about 100 mg/kg/day in chickens given 1000 ppm 2,4-D or 2,4,5-T in their drinking water for up to 2 years (Erne, 1966a; Bjorklund and Erne, 1971). These estimated values are based on a water intake of 1 ml/hr for a rat weighing 200 g, 600 ml/hr for a calf weighing 300 kg, and 7.5 ml/hr in a chicken weighing 2 kg (Spector, 1956).

TABLE 2

*Dosage levels in animals given multiple doses of phenoxy herbicides (mg/kg/day for x days).*

| Animals | 2,4-D | MCPA | 2,4,5-T | Silvex |
|---|---|---|---|---|
| Rats | ≈120 (60x)[1] | | 50 (5x)[3] | |
| Cows | 0.3 (5x)[4]<br>3 (4x)[2]<br>0.9 - 30<br>(14 - 21x)[5, 6] | 3 (4x)[2]<br>3 - 30<br>(14 - 21x)[5, 6] | 0.3 (4x)[7]<br>0.03 - 30<br>(14 - 21x)[5, 6] | 0.3 (4x)[7]<br>3 - 30<br>(14 - 21x)[5, 6] |
| Calves | 8 - 60<br>(28x)[6, 8, 9]<br>≈ 50 (>60x)[1] | 7 - 15<br>(28x)[6, 8] | 3.4 - 63<br>(28x)[6, 8]<br>0.15, 0.75<br>(up to 224x)[10] | 8 - 54<br>(28x)[6, 8, 9] |
| Sheep | 55 - 62<br>(28x)[6, 8, 9] | 15 (28x)[6, 8] | 52 - 60<br>(28x)[6, 8, 9]<br>0.15, 0.75,<br>250 (4x)[10] | 33 - 58<br>(28x)[6, 8, 9] |
| Pigs | 50 (3 - 30x)[1]<br>≈ 20 (60 - 180x)[1] | | | |
| Chickens | 300 (12 - 24x)[1]<br>≈ 100 (60 - 730x)[1]<br>≈ 100 (14 - 201x)[11]<br>0.7, 7 (30x)[12] | | ≈100 (14 - 201x)[11] | |

[1] Erne, 1966a
[2] Bache, 1964a
[3] Courtney, 1970
[4] Gutenmann, 1963a
[5] Bjerke, 1971, 1972
[6] Leng, 1972
[7] St. John, 1964
[8] Jensen, 1971, 1973
[9] Clark, 1971a, b; 1975
[10] Clark, 1971c
[11] Bjorklund, 1971
[12] Dow, unpublished

*Toxicity of Phenoxy Herbicides in Animals*

Many of the studies listed in Tables 1 and 2 were conducted at dosage levels likely to cause symptoms of toxicity in the animals. For example, the acute oral $LD_{50}$ of phenoxy herbicides is in the range of 300 to 1000 mg/kg for laboratory and domestic animals (Palmer et al., 1969). Generally, dogs were more sensitive to 2,4-D and 2,4,5-T on a weight basis, and chicks were more tolerant. Pigs were more sensitive than calves in comparable studies at 50 and 100 mg/kg/day (Erne, 1966a). Dosage levels which caused no ill-effects ranged from about 2 to 60 mg/kg/day in subacute and chronic feeding studies with these herbicides in rats and dogs (Mullison, 1966; Way, 1969; Hansen et al., 1971; WSSA, 1974; Dow, unpublished data).

It should be noted that dosage levels listed in Table 2 cover a 10,000-fold range, from 0.03 mg/kg/day in cows to 300 mg/kg/day in chickens, for 2 to 3 weeks in each species.

*Metabolic Pathways*

Proposed pathways for metabolism of 2,4-D are depicted in Fig. 2. Phenoxy herbicide products are generally formulated as alkali salts, amine salts, and esters ranging from highly volatile isopropyl or butyl esters to low-volatile polypropyleneglycol butyl ether esters. Such salts and esters are rapidly hydrolyzed in plants and animals. The phenoxy acids so formed exist mainly as organic anions at the pH of living tissue, and may form conjugates with amino acids. Such conjugates are readily dissociated by treatment with acid or base since conjugation is through the amino group rather than the carboxyl group of amino acids (or proteins). Similarly, conjugation of phenoxy acids with glucuronic acid is via reaction with a hydroxyl group rather than the carboxyl group of the glucuronic acid, so the conjugates are phenoxy esters rather than glucuronates.

The ether link of phenoxyalkanoic acids may also be cleaved to release the corresponding phenols, chiefly by microbial action. Such phenols have been found in soil (Loos, 1975) and in tissues of ruminants, but seldom in plants or even in urine of other animals, as discussed below. The rate of cleavage of the ether link is hindered by substitution of an angular methyl group on the alkyl side chain as in silvex, or a chlorine in the meta position on the ring as in 2,4,5-T and silvex.

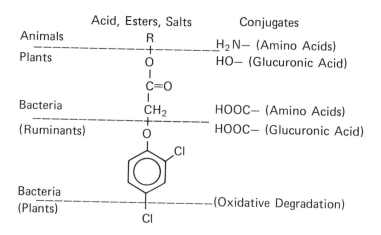

Fig. 2. *Pathways for metabolism of phenoxy herbicides.*

Oxidative degradation of the ring may take place, probably by initial attack at the 4 position, again chiefly through bacterial action. However, it is not likely to occur during the short residence time of phenoxy herbicides in the body of animals. Oxidation is also not likely to occur in the rumen where conditions are more conducive to reduction (redox potential of -0.35 volts).

General observations and conclusions drawn from the animal studies listed in Tables 1 and 2 can be summarized as follows:

1. Phenoxy acids, salts, and esters are readily absorbed from the gut and become widely distributed in the tissues of animals (Erne, 1966a,b; Courtney, 1970; Fang et al., 1973; Lindquist, 1974; Sauerhoff et al., 1976a,c; Dow, unpublished data).

2. Peak plasma concentrations are achieved within 2 to 24 hours after administration, depending on species and dosage level (Erne, 1966a; Piper et al., 1973), and route of administration (Sauerhoff et al., 1976a,c).

3. Phenoxy esters are rapidly hydrolyzed to the corresponding acids in animals (Erne, 1966a; Clark, 1969; St. John et al., 1964; Dow, unpublished data).

4. Phenoxy acids are reversibly bound to plasma proteins (Erne, 1966b; Gehring, et al., 1973).

5. Peak plasma levels may be lower after repeated doses than after the first dose, indicating development of tolerance by some individuals or in some species (Erne, 1966a).

6. Phenoxy acids are excreted rapidly in the urine as unchanged acids or as conjugates which can be hydrolyzed readily with acid to give the parent compound (Erne, 1966b; Grunow, 1971; Shafik et al., 1971; Piper et al., 1973; Dow, unpublished data).

7. Relatively quantitative recovery is obtained in urine within less than one week after a single dose, depending on dose level and on species (Erne, 1966a; Shafik et al., 1971; Piper et al., 1973; Gehring et al., 1973; Sauerhoff et al., 1976a,b,c).

8. Part of the administered dose may be excreted in feces, particularly in dogs (Piper et al., 1973; Dow, unpublished data).

9. Some metabolism may occur at high dosage rates in rats (Grunow et al., 1971; Shafik et al., 1971; Piper et al., 1973), in dogs (Piper et al., 1973), and in cattle and sheep (Leng, 1972).

10. The rate of clearance from the body is dependent on the dose level, particularly if the renal threshold is exceeded (Piper et al., 1973; Hook et al., 1974; Sauerhoff et al., 1976a,c).

11. Only minor differences in pattern of distribution and elimination were noted after intravenous vs. oral administration of $^{14}$C-labeled 2,4,5-T and silvex as salts or esters in rats (Sauerhoff, 1976a,c; Dow, unpublished data).

12. Cleavage of the phenoxy ether link occurs in ruminants, resulting in residues of phenol metabolites in milk, liver, and kidney of animals fed high doses of the herbicides (Bjerke et al., 1972; Leng, 1972; Jensen et al., 1973; Clark et al., 1975). Rates of formation and degradation or elimination of the phenol metabolites from the body depend on dosage levels and on chemical structure of the phenoxy compounds fed to the animals. Only traces of phenol metabolites were detected in urine of rats, primarily when high doses of 2,4,5-T were administered (Shafik et al., 1971; Piper et al., 1973), and none was detected in urine of humans (Shafik et al., 1971; Gehring et al., 1973; Dow, unpublished data).

## PHARMACOKINETIC STUDIES

Data from some of the more extensive studies serve to illustrate why different results can be obtained in apparently similar experiments, depending on dose levels, routes of administration, duration of studies, species of animals, and structure of the phenoxy herbicide given to the animals.

### Effect of Dosage Level

In pharmacokinetic/metabolism studies by scientists of The Dow Chemical Company, groups of three male and three female rats were given $^{14}$C-carboxyl-labeled 2,4,5-T as a single oral dose at 5, 50, 100 or 200 mg/kg (Piper et al., 1973). As shown in Table 3, the peak plasma concentration and volume of distribution increased with increasing dose.

The volume of distribution is a calculated value developed by computer analysis of plasma data using equations for one or two compartment models, depending on best fit for the data. The increasing volume of distribution with dosage level indicates that more of the 2,4,5-T was finding its way into tissues and cells following administration of higher doses.

TABLE 3

*Effect of dosage level on fate of 2,4,5-T in rats.*

| | | | | |
|---|---|---|---|---|
| Single Oral Dose, mg/kg | 5 | 50 | 100 | 200 |
| Number of Animals (M + F) | 6 | 6 | 6 | 6 |
| Duration of Study, Days | 4 | 5 | 5 | 6 |
| Peak Plasma Conc., ug/ml | 15 | 90 | 150 | 250 |
| Interval Post-Administration, Hour | 12 | 12 | 24 | 24 |
| Volume of Distribution, ml/kg | 144 | 191 | 385 | 437 |
| Number of Body Compartments | 1 | 1 | 1 | 1 |
| Average Rate of Clearance, $T_{½}$, Hour | | | | |
|   from Plasma | 4.7 | 4.2 | 19.4 | 25.2 |
|   from Body via Urine | 13.6 | 13.1 | 19.3 | 28.9 |
| Excretion as Percent of Dose | | | | |
|   in Urine | 83 | 93 | 78 | 68 |
|   (as metabolites) | — | — | ($<$5) | (5) |
|   in Feces | 3 | 6 | 9 | 14 |
| Total in Excreta as Percent of Dose | 86 | 99 | 87 | 82 |

[1]Piper et al., 1973

Of particular interest is the difference in pattern of clearance of the two lower dosage levels compared to the two higher levels. At dosage levels of 5 and 50 mg/kg, the rates of clearance were 4.7 and 4.2 hr for plasma compared to 13.6 and 13.1 hr for clearance in the urine. This more rapid elimination from plasma indicates active uptake of 2,4,5-T by the kidney in rats at these dosage levels. However, at dosage levels of 100 or 200 mg/kg, the rates of clearance from plasma and the body were similar indicating that active uptake by the kidney had been saturated at these dosages.

Most of the administered dose was excreted in urine within 36 hr after administration of 5 or 50 mg/kg. More time was needed to achieve total recovery of higher doses and proportionally more was excreted in feces. All the 2,4,5-T in urine was accounted for as the free acid by the method of analysis used in the Dow study, except for traces of metabolites at the higher doses. On the other hand, German workers reported recovery of only 45 to 70% of a 50 mg/kg dose in urine of rats in 7 days (Grunow et al., 1971). They also reported finding up to one-third of the 2,4,5-T in urine as a glycine conjugate which could be converted to 2,4,5-T by acid hydrolysis. The glycine conjugate was isolated and identified in urine of rats given a single oral dose of 75 mg 2,4,5-T. (It should be noted that a dose of 75 mg is equivalent to 300 mg/kg in a 250 g rat.)

*Fate of 2,4,5-T in Rats, Dogs, and Humans*

Dow scientists also conducted studies with 2,4,5-T at 5 mg/kg in dogs and humans (Piper et al., 1973; Gehring et al., 1973). A comparison of pharmacokinetic data for rats, dogs and humans is presented in Table 4.

TABLE 4

*Effect of species on fate of 2,4,5-T in animals.*

| Species Studied | Rat[1] | Dog[1] | Human[2] |
|---|---|---|---|
| Single Oral Dose, mg/kg | 5 | 5 | 5 |
| Number of Animals/Tests | 6 | 4 | 5/7 |
| Duration of Study, Days | 4-6 | 9 | 4 |
| Peak Plasma Conc., ug/ml | 15 | 20 | 57 |
| Interval Post-Administration, Hour | 12 | $<4$ | 7 |
| Volume of Distribution, ml/kg | 144 | 221 | 80 |
| Number of Body Compartments | 1 | 1 | 1 |
| Average Rate of Clearance, $T_{½}$, Hour | | | |
|   from Plasma | 4.7 | 77 | 23.1 |
|   from Body via Urine | 13.6 | 87 | 23.1 |
| Excretion as Percent of Dose | | | |
|   in Urine, Total Ether Soluble | 83 | 42 | 89 |
|   (as Altered 2,4,5-T) | (none) | (4) | (little) |
|   in Feces | (little) | 20 | $<1$ |
| Total in Excreta as Percent of Dose | 83 | 62 | 90 |

[1] Piper et al., 1973.
[2] Gehring et al., 1973.

The peak plasma concentration was lower in rats and dogs than in humans. Analogously, the volume of distribution was higher in rats and dogs than in humans, indicating that 2,4,5-T was distributed more diffusely in the body of rats and dogs than in humans. Most of the 2,4,5-T residues were in the plasma, particularly in man, as indicated by the high plasma level and low volume of distribution. Dialysis studies demonstrated that virtually all of the 2,4,5-T in plasma is "bound" reversibly to protein.

Peak plasma concentrations were achieved more rapidly in dogs and humans than in rats. However, rates of clearance from plasma and body were considerably slower in dogs than in rats or humans, probably due to the poor secretory process for organic acids in the dog kidney (Hook et al., 1974).

Relatively complete recovery of the administered dose was obtained as unchanged 2,4,5-T in urine of rats and humans within 4 to 6 days after dosing. In dogs, on the other hand, only 62% of the administered dose was excreted in 9 days, and one-third of the amount excreted was found in feces, partly as altered 2,4,5-T. This illustrates a pertinent principle if elimination via one route (urine) is delayed, a greater fraction will be eliminated via other routes (feces).

## Fate of 2,4-D, 2,4,5-T, and Silvex in Humans

Dow scientists also conducted studies in humans given oral doses of 2,4-D at 5 mg/kg (Sauerhoff, 1976b) and silvex at 1 mg/kg (unpublished data). As shown in Table 5, peak plasma concentrations were proportionally lower and were achieved more rapidly for 2,4-D and silvex than for 2,4,5-T. Computer analysis of data for plasma levels indicated that a two-compartment model provided a better fit for 2,4-D in one subject and for silvex as the free acid in all subjects, in contrast to the one-compartment model found for 2,4,5-T.

Rates of clearance were faster for 2,4-D and silvex than for 2,4,5-T from plasma and the body in humans. Furthermore, excretion of silvex appeared to be biphasic, with a half-life of less than 5 hours for the first 2 days, and of 26 hours for the next 4 days.

Relatively complete recovery of the administered dose was obtained for all three chemicals in the urine of a total of 18 human subjects. However, up to 57% of the silvex excreted in urine was in conjugated form, compared to 0 to 27% for 2,4-D and trace amounts for 2,4,5-T. Acid hydrolysis of urine or an ether extract of urine liberated unchanged silvex from the conjugate, indicating that silvex may exist in association with amino acid(s), at least after passage through the kidney.

TABLE 5

Fate of three phenoxy herbicides in humans.

| Phenoxy Herbicide Administered | 2,4,5-T[1] | 2,4-D[2] | Silvex[3] |
|---|---|---|---|
| Single Oral Dose, mg/kg | 5 | 5 | 1 |
| Number of Subjects/Tests | 5/7 | 5 | 8 |
| Duration of Study, Days | 4 - 6 | 4 | 6 |
| Peak Plasma Concentration, ug/ml | 57 | 25 | 6 |
| Interval Post Administration, Hour | 7 | 4 | 2 - 4 |
| Volume of Distribution, ml/kg | 80 | >200, 83 | 115, 107 |
| Number of Body Compartments | 1 | 1 or 2 | 2 |
| Average Rate of Clearance, $T_{\frac{1}{2}}$, Hour | | | |
|   from Plasma | 23.1 | 11.7 | 3.7, 19 |
|   from Body via Urine | 23.1 | 17.7 | 5, 26 |
| Excretion as Percent of Dose | | | |
|   in Urine as Free Acid | 88.5±5.1 | 70 - 88 | 30 - 80 |
|   as Conjugate(s) | (little) | 0 - 27 | 15 - 54 |
|   in Feces (First 2 Days) | < 1 | — | 0 - 3 |
| Total Excreted as Percent of Dose | ≈90 | 88 - 106 | 67 - 95 |

[1] Gehring et al., 1973
[2] Sauerhoff et al., 1976b
[3] Dow, Unpublished

No phenol metabolites were detected in urine of rats, dogs, or humans after single oral doses of 2,4,5-T, 2,4-D, or silvex at 5 or 1 mg/kg of body weight.

## RESIDUES IN ANIMAL TISSUES

Thousands of residue analyses were conducted by Dow chemists in obtaining data necessary for establishment of tolerances for phenoxy herbicides and their corresponding phenol metabolites in milk and meat. Reports of the studies were presented at meetings of the American Chemical Society in Washington and Dallas (Bjerke et al., 1971; Jensen et al., 1971, 1973), and summaries of some of the Dow data have been published (Bjerke et al., 1972; Leng, 1972). Some of the meat samples were also analyzed by USDA chemists who reported similar findings (Clark et al., 1971a,b; 1975).

Results of the Dow analyses are portrayed as bar graphs in Figures 3 through 9. Each bar represents the average residue of phenoxy herbicide or phenol metabolite found in samples of milk, muscle, fat, liver, or kidney from three cows or three beef calves at each feeding level. The values include "bound" or conjugated residues released by acid and/or alkaline hydrolysis of the samples (Jensen et al., 1973).

*Residues in Milk and Cream*

Fig. 3 shows that no residues of 2,4-D, MCPA or silvex were detected in milk of cows fed 100 or 300 ppm phenoxy herbicide in the diet for 14 days. Only very low levels (<0.1 ppm) were found when the animals were maintained on feed containing 1000 ppm herbicide, equivalent to a dosage level of about 30 mg/kg/day in the cows. On the other hand, traces of 2,4,5-T were detected in milk of cows fed 300 ppm in the diet and the residues increased to 0.4 ppm at 1000 ppm

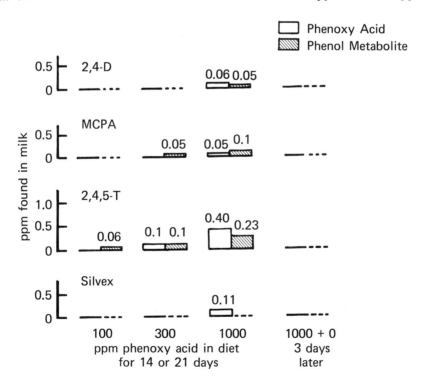

Fig. 3. Residues in milk of cows fed phenoxy acids.

in the diet. Such high levels in feed might occur at the time of spraying pastures at high rates to control brush but the residue in forage would decline rapidly with a half-life of only 1 to 2 weeks (Leng, 1972).

Trace residues of phenolic metabolites were also detected in milk of cows maintained on feed containing 1000 ppm 2,4-D, 300 ppm MCPA, or 100 ppm 2,4,5-T, but not in any samples from cows fed silvex. All residues declined to non-detectable levels in milk within 3 days after withdrawal of the cows from 1000 ppm phenoxy herbicide in the total diet for 21 days. Corresponding cream samples contained comparable residues of phenoxy herbicides and phenol metabolites, indicating that these residues do not concentrate in fat.

*Residues of 2,4-D in Meat*

Fig. 4 represents the residue pattern found for 2,4-D and 2,4-dichlorophenol in muscle, fat, liver, and kidney of calves fed 300, 1000 or 2000 ppm 2,4-D in the diet for 28 days before slaughter, and for animals fed untreated feed for 7 days after withdrawal from the highest feeding level. These feeding levels corresponded to about 9, 30, and 60 mg/kg/day for the three animals at each level.

No residues were detected in muscle and only traces of 2,4-D were found in fat at each feeding level. Residues of 2,4-D and its phenol metabolite were found in liver of animals fed 1000 or 2000 ppm 2,4-D. Much higher residues were found in kidney, including 0.4 to 1.8 ppm of the phenol metabolite. Residues of 2,4-D declined rapidly after withdrawal of the animals from feed containing 2000 ppm 2,4-D. Less than 0.2 ppm of the phenol metabolite remained in liver and kidney at 7 days after discontinuation of treatment at this exaggerated level in the feed.

Residues were similar in tissues of sheep fed 2000 ppm 2,4-D except for much lower levels of the phenol metabolite in kidney of sheep.

*Residues of MCPA in Meat*

Fig. 5 presents data for residues in tissues of calves fed 250 or 500 ppm MCPA for 28 days before slaughter. The residues found were similar to those for 2,4-D at 300 ppm in the diet but no phenol metabolite was detected even at 500 ppm MCPA in the diet. Residues in sheep were generally lower than in calves fed 500 ppm MCPA.

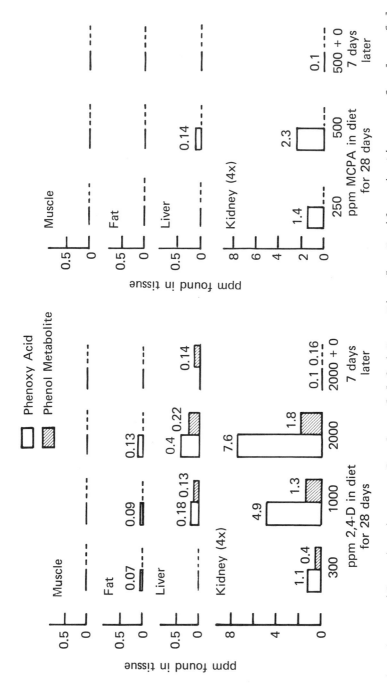

Fig. 4. Residues in tissues of calves fed 2,4-D.

Fig. 5. Residues in tissues of calves fed MCPA.

## Residues of 2,4,5-T in Meat

Fig. 6 demonstrates that low levels of 2,4,5-T occurred in muscle and fat of calves given 300 or 900 ppm in the diet, and that much higher residues were found in tissues of animals fed 1800 ppm for 28 days. Only a trace of 2,4,5-trichlorophenol was found in fat at the highest feeding level. All residues disappeared from muscle and fat within 7 days after withdrawal from this exaggerated feeding level.

Residues in liver[1] were considerably higher for 2,4,5-T than for 2,4-D. As noted for muscle and fat, residues of 2,4,5-T were relatively low in liver at feeding levels of 300 and 900 ppm, but increased sharply for 1800 ppm in the feed. This indicates that the threshold level may have been exceeded at this dosage level. On the other hand, residues of 2,4,5-T in kidney[1] appeared to be proportional to the level in the diet. Residues of 2,4,5-trichlorophenol metabolite were proportionally higher in liver than in kidney, and declined more slowly after withdrawal of the animals from a diet containing 1800 ppm 2,4,5-T.

Residues in tissues of sheep fed 2000 ppm 2,4,5-T were considerably lower than those in beef animals fed 1800 ppm, except for trichlorophenol metabolite in liver.

## Residues of Silvex in Meat

Fig. 7 demonstrates that residues of silvex in muscle and fat were proportional to levels in the diet and decreased to less than 10% of the initial level within 7 days after withdrawal from 2000 ppm in the diet. The same pattern was obtained for high residues in liver and kidney.[1] Only traces of 2,4,5-trichlorophenol were found in liver of calves fed 2000 ppm silvex, and none was detected in kidney. On the other hand, 2 to 3 ppm of the same trichlorophenol metabolite was found in liver and kidney of calves fed 1800 ppm 2,4,5-T. This indicates that the angular methyl group on the side chain in silvex hinders cleavage of the phenoxy ether link through action of rumen bacteria.

Residues in tissues of sheep fed 2000 ppm silvex were about half the levels found in tissues of calves at the same feeding level.

---

[1] Note eight-fold change in scale for residues in liver and kidney compared to the scale for residues in muscle and fat. A four-fold change in scale is also used for residues of 2,4-D and MCPA in kidney (Fig. 4 and Fig. 5).

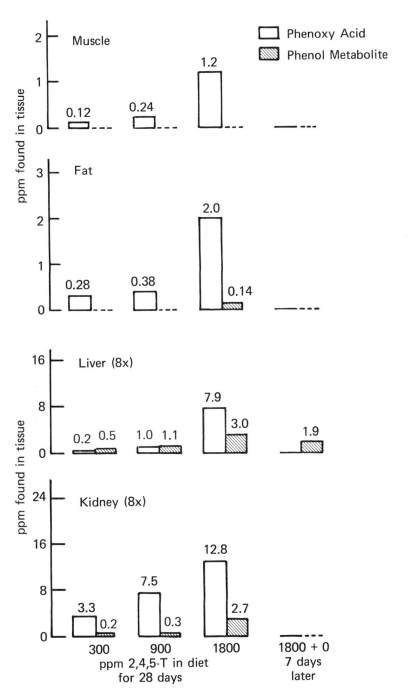

Fig. 6. Residues in tissues of calves fed 2,4,5-T.

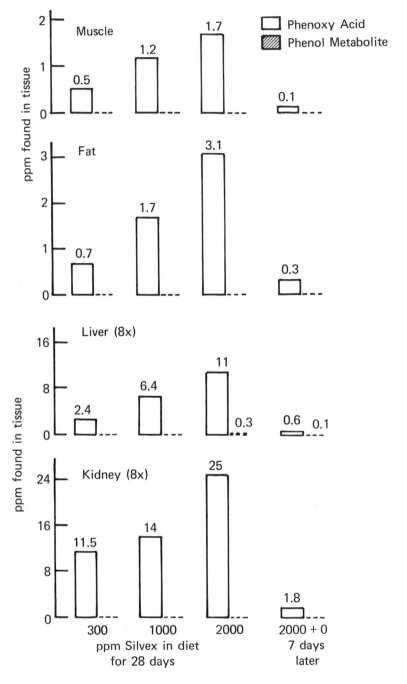

Fig. 7. Residues in tissues of calves fed silvex.

## Comparison of Residues in Liver and Kidney

Figs. 8 and 9 illustrate the difference in metabolic pattern for the four phenoxy herbicides at low and high levels in the diet of calves. High levels of phenol metabolite were found in liver only for animals fed high levels of 2,4,5-T. Release of most of the 2,4,5-trichlorophenol metabolite was accomplished only after alkaline digestion of the tissue, indicating that phenol metabolites may be more tightly "bound" or conjugated in liver than in other tissues.

Fig. 9 compares residues found in kidney under the same conditions as in Fig. 8. Residues of phenol metabolite were found only in samples from animals fed 2000 ppm 2,4-D or 1800 ppm 2,4,5-T. The phenol metabolites in kidney were extractable with ether/hexane, in contrast to unextractable residues in liver.

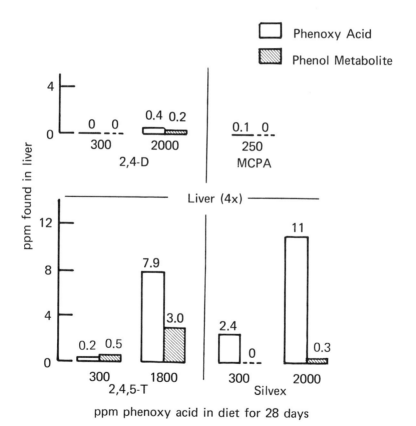

*Fig. 8. Comparison of residues in liver of calves fed two levels of four phenoxy acids.*

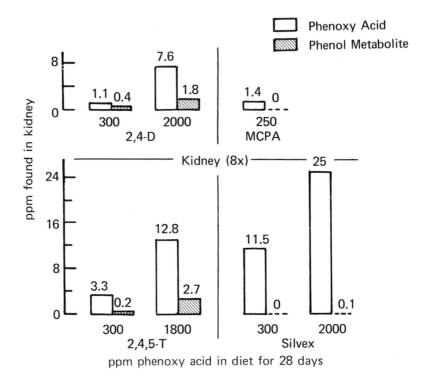

Fig. 9. Comparison of residues in kidney of calves fed two levels of four phenoxy acids.

DISCUSSION

The fate of phenoxy herbicides in animals depends on a number of factors, including differences in chemical structure and differences among animal species. Also of major importance are the dosage levels given to the animals, and to a lesser extent the routes of administration and duration of studies. The following pictorial illustration demonstrates the fallacy of conducting metabolism studies at highly exaggerated dosage levels which exceed the threshold for normal handling of the chemical being investigated.

Fig. 10 portrays a barrel containing two slits. The first slit represents normal metabolic pathways through which small doses are eliminated. An example might be clearance of 2,4,5-T from the body of rats by active uptake in the kidney for dosage levels of 5 or 50 mg/kg, depicted by cases A and B. The other slit represents a secondary pathway which operates only when the system is overloaded, as in cases C and D. Such a pathway may account for slower clearance of 2,4,5-T from plasma of rats given dosage levels of 100 or 200 mg/kg, with consequent slower overall rate of elimination from the body at these higher dosage levels. It could also account for the appearance of metabolites in urine, or excretion in feces, at the higher dosage levels.

The size, distribution and number of slits in the barrel may depend on the species of animal, as observed for elimination of 2,4,5-T in dogs compared to rats or humans. In this case, humans resembled rats more than dogs and would likely tolerate dosage levels of 2,4,5-T which might be toxic in dogs.

The slit barrel concept might also explain differences observed in residue patterns for 2,4-D, MCPA, 2,4,5-T, and silvex in milk, muscle, fat, liver, and kidney of cattle fed comparable but exaggerated levels of these herbicides in the diet.

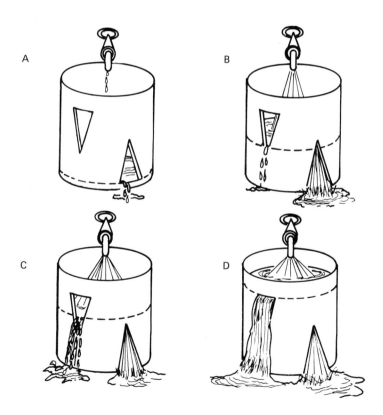

Fig. 10. Effect of dosage level on metabolic pathway (Gehring et al., 1976).

## CONCLUSIONS

Residues of phenoxy alkanoic acids in treated food or feed crops are absorbed readily in the gut of animals and are excreted rapidly in the urine, largely as unchanged phenoxy acid. Some conjugation may occur, particularly at higher dosage levels, but the basic structure of the herbicide is not readily altered in animals. The ether linkage can be cleaved by bacterial action in the rumen but the rate of cleavage depends on the chemical structure of the phenoxy compound. The rate of clearance of residues from the body is dependent on dosage level, particularly if the renal threshold is exceeded.

Residues of the phenoxy acids, and in some cases the corresponding phenol moiety, are dependent on the dosage level and on the chemical structure of the herbicide in the feed of animals. Residues are considerably lower in muscle, fat, milk, and cream than in liver and kidney, but all residues decline rapidly after withdrawal of animals from treated feed. Residues of phenol metabolites may be present in milk, liver, and kidney of animals fed high doses of 2,4-D, MCPA, and 2,4,5-T, but not of silvex.

Results of studies at high dosage levels cannot be extrapolated to what is likely to happen at low dosage levels representative of what might be encountered from ingestion of treated foods or feeds. Humans resembled rats rather than dogs in comparable studies with 2,4,5-T indicating that the dog is not a suitable model for predicting the fate of phenoxy herbicides in humans.

## ACKNOWLEDGEMENT

The author wishes to thank P. J. Gehring and members of his research group for permission to include unpublished data on pharmacokinetic studies with phenoxy herbicides in animals and humans, and D. J. Jensen for unpublished data on residues in beef tissues.

## REFERENCES

Bache, C. A., Hardee, D. D., Holland, R. F., Lisk, D. J., J. Dairy Sci. 47, 298 (1964a).
Bache, C. A., Lisk, D. J., Wagner, D. G., Warner, R. G., J. Dairy Sci. 47, 93 (1964b).
Bjerke, E. L., Herman, J. L., Miller, P. W., Wetters, J. H., presented at the 162nd National Meeting of the American Chemical Society, Washington DC, PEST-57 (September 1971).

Bjerke, E. L., Herman, J. L., Miller, P. W., Wetters, J. H., J. Agric. Food Chem. 20, 963 (1972).
Bjorklund, N.-E., Erne, K., Acta Vet. Scand. 12, 243 (1971).
Clark, D. E., J. Agric. Food Chem. 17, 1168 (1969).
Clark, D. E., Crookshank, H. R., Radeleff, R. D., Palmer, J. S., presented at the 161st National Meeting of the American Chemical Society, Los Angeles CA, PEST-10 (March 1971a).
Clark, D. E., Crookshank, H. R., Radeleff, R. D., Palmer, J. S., Binns, W., presented at the 162nd National Meeting of the American Chemical Society, Washington DC, PEST-32 (September 1971b).
Clark, D. E., Palmer, J. S., J. Agric. Food Chem. 19, 761 (1971c).
Clark, D. E., Palmer, J. S., Radeleff, R. D., Crookshank, H. R., Farr, F. M., J. Agric. Food Chem. 23, 573 (1975).
Clark, D. E., Young, J. E., Younger, R. L., Hunt, L. M., McLaran, J. K., J. Agric. Food Chem. 12, 43 (1964).
Courtney, D. K., in "Pesticides Symposia," Deichmann, W. B., Ed., University of Miami School of Medicine, Miami FL, 1970, p. 277.
Environmental Protection Agency, 40 CFR §180.399 (MCPA), Federal Register 40, 51045 (November 3, 1975).
Environmental Protection Agency, 40 CFR §180.142) 2,4-D), Federal Register 41, 11514 (March 19, 1976).
Erne, K., Acta Vet. Scand. 7, 240 (1966a).
Erne, K., Acta Vet. Scand. 7, 264 (1966b).
Fang, S. C., Fallin, E., Montgomery, M. L., Freed, V. H., Toxicol. Appl. Pharmacol. 24, 555 (1973).
Gehring, P. J., Blau, G. E., Watanabe, P. G., in "Advances in Modern Toxicology," Hemisphere Pub. Corp., Washington DC, 1976.
Gehring, P. J., Kramer, C. G., Schwetz, B. A., Rose, J. Q., Rowe, V. K., Toxicol. Appl. Pharmacol. 26, 352 (1973).
Grunow, W., Bome, C., Budczies, B., Food Cosmet. Toxicol. 9, 667 (1971).
Gutenmann, W. H., Hardee, D. D., Holland, R. F., Lisk, D. J., J. Dairy Sci. 46, 991 (1963a).
Gutenmann, W. H., Hardee, D. D., Holland, R. F., Lisk, D. J., J. Dairy Sci. 46, 1287 (1963b).
Hansen, W. H., Quaife, M. L., Haberman, R. T., Fitzhugh, O. G., Toxicol. Appl. Pharmacol. 20, 122 (1971).
Hook, J. B., Bailie, M. D., Johnson, J. T., Gehring, P. J., Food Cosmet. Toxicol. 12, 209 (1974).
Jensen, D. J., Bjerke, E. L., Herman, J. L., Miller, P. W., Berkenke, L. F., presented at the 162nd National Meeting of the American Chemical Society, Washington DC, PEST-56 (September 1971).

Jensen, D. J., Miller, P. W., Berhenke, L. F., presented at the 165th National Meeting of the American Chemical Society, Dallas TX, PEST-22 (April 1973).
Khanna, S., Fang, S. C., *J. Agric. Food Chem.* 14, 500 (1966).
Kohli, J. D., Khanna, R. N., Gupta, B. N., *Arch. Int. Pharmacodyn. Ther.* 210, 250 (1974).
Leng, M. L., *DOWN TO EARTH* 28, 12 (Summer 1972).
Lindquist, N. G., *Toxicol. Appl. Pharmacol.* 30, 227 (1974).
Lisk, D. J., Gutenmann, W. H., Bache, C. A., Warner, R. G., Wagner, D. G., *J. Dairy Sci.* 46, 1435 (1963).
Loos, M. A., in "Herbicides: Chemistry, Degradation, and Mode of Action," Kearney, P. C., Kaufmann, D. D., Ec., 2nd ed, Vol. 1, Marcel Dekker, Inc., New York NY, 1975.
Matsumura, A., *Japanese J. Ind. Health* 12, 20 (1970).
Mullison, W. R., presented at Southern Weed Conference, Jacksonville FL, January 1966.
Palmer, J. S., Radeleff, R. D., "Production Research Report No. 106," ARS, USDA, Washington DC, 1969.
Paulson, G. D., *Residue Rev.* 58, 1 (1975).
Piper, W. N., Rose, J. Q., Leng, M. L., Gehring, P. J., *Toxicol. Appl. Pharmacol.* 26, 339 (1973).
Shafik, M. T., Sullivan, H. C., Enos, H. R., *Intern. J. Environ. Anal. Chem.* 1, 23 (1971).
Sauerhoff, M. W., Braun, W. H., Blau, G. E., Gehring, P. J., *Toxicol. Appl. Pharmacol.* 36, 491 (1976a).
Sauerhoff, M. W., Braun, W. H., Blau, G. E., LeBeau, J. E., presented at the 15th Annual Meeting of the Society of Toxicology, Atlanta GA, No. 107, (March 1976b).
Sauerhoff, M. W., Braun, W. H., Gehring, P. J., presented at the 15th Annual Meeting of the Society of Toxicology, Atlanta GA, No. 6 (March 1976c).
Spector, W. S., "Handbook of Biological Data," W. B. Saunders Co., Philadelphia, 1956, p. 355.
St. John, L. E. Jr., Wagner, D. G., Lisk, D. J., *J. Dairy Sci.* 47, 1267 (1964).
Way, J. M., *Residue Rev.* 26, 37 (1969).
WSSA "Herbicide Handbook," 3rd ed, 1974, p. 116, 232, 359, 375.
Zielinski, W. L., Fishbein, L., *J. Agric. Food Chem.* 15, 841 (1967).

COMPARATIVE METABOLISM AND AN EXPERIMENTAL
APPROACH FOR STUDY OF LIVER OXIDASE
INDUCTION IN PRIMATES


R. I. Krieger, J. L. Miller, S. J. Gee,
and C. R. Clark

*Department of Environmental Toxicology
University of California, Davis*


*ABSTRACT. Although there is a close systematic relationship between apes, man and monkeys, these groups are infrequently used in insecticide disposition studies. The need to know about insecticide disposition derives from predictable or unintended insecticide exposures in the personal, occupational, and community environments. Subhuman primates may be particularly well suited to investigations of target organ toxicity following subacute and chronic insecticide exposures. Usefulness of these studies for assessing potential harm to man will depend upon knowledge of absorption, distribution, biotransformation, deposition, macromolecular binding, and elimination processes. Representatives of major insecticide classes have been studied in primates, but knowledge of the fate of most insecticides can only be inferred. Recent comparative studies in this laboratory using rhesus monkeys, Macaca mulata, have measured the effects of liver monooxygenase inducers on liver ultrastructure, antipyrine half-life in vivo and oxidative metabolism in vitro of aldrin, dihydroisodrin, p-nitroanisole, and 3,4-benzpyrene. Closed liver needle biopsies were the enzyme source for in vitro work. Reduced antipyrine plasma half-lives and increased in vitro levels of aldrin epoxidation and p-nitroanisole O-demethylation were obtained after feeding 10, 100, and 500 ppm DDT in diet for consecutive 30 day periods. Induction was also obtained with phenobarbital (15 mg/kg, 2 x/day x 4, im). Animals served as their own controls and showed no overt or clinical signs of toxicity. Thus, monkeys can be used in comparative disposition studies in which knowledge of the fate of xenobiotics is essential for evaluation of their biological activity.*

## INTRODUCTION

Humans are included among the 202 species of living primates (Napier and Napier, 1967). The use of primates in predictive toxicology is based upon this close phylogenetic relationship. Commonly the rhesus monkey *Macaca mulatta*, stumptail monkey *M. arctoides*, Japanese monkey *M. fuscata*, squirrel monkey *Saimiri sciuresus* and the chimpanzee *Pan troglodytes* are used in preclinical evaluations of drugs. Use of primates in biomedical research, particularly drug metabolism, has been discussed in the following publications: Conference on Nonhuman Primate Toxicology (Miller, 1968), Coulston (1966), Coulston and Serrone (1969), Drug Research Board (1969), Heywood (1972), Reulis (1975), Smith (1966, 1972), Smith and Williams (1974), Udall (1969) and Williams (1971, 1976). Although these reviews describe primarily work with drugs they are appropriate for consideration in discussions of foreign compound metabolism in general since metabolic processes are operative irrespective of a chemical's particular pattern of use.

Whereas drugs are used as curatives of disease and human exposure is predictable, insecticides are intended for pest control so exposures are usually unintentional. Accordingly, preregistration testing of insecticides does not routinely include metabolism studies with primates, hence experimental data are extremely limited. The Report of the Secretary's Commission on Pesticides and Their Relationship to Environmental Health (USDHEW, 1969) includes a review of effects of pesticides on humans. The base of knowledge is derived from the following: 1) limited controlled administrations of insecticides to human subjects, 2) case reports of episodes of accidental poisoning, and 3) epidemiological studies of occupationally-exposed groups and the general population. Appropriately there is a strong clinical emphasis in most of this work, so studies of metabolism of the test compound are generally incomplete.

This paper will primarily discuss experimental, rather than epidemiological, studies of the fate and metabolic effects of foreign compounds including insecticides in primates. The status and use of subhuman primates in biomedical research will be briefly reviewed. Some comprehensive comparative studies that have revealed metabolic similarities between humans and other primates will be discussed. Finally, chlorinated hydrocarbon insecticide induction studies in primates, including occupationally-exposed humans, will be noted in relation to current studies of methods for continually assessing liver enzyme induction in rhesus monkeys.

TABLE 1

*Partial summary of survey of vertebrates used for research during 1971*[1]

| | | |
|---|---|---:|
| Mice | | 30,281,783 |
| Rats | | 10,204,069 |
| Frogs | | 3,315,677 |
| Chickens | | 3,042,968 |
| Hamsters | | 938,065 |
| Guinea Pigs | | 646,322 |
| Rabbits | | 442,614 |
| Swine | | 208,346 |
| Dogs | | 192,524 |
| Turtles | | 103,721 |
| Subhuman primates | | 56,567 |
| Gerbils | | 49,541 |
| Cattle | | 22,490 |
| Goats | | 3,140 |
| | TOTAL | 49,546,983 |

[1]National Research Council, 1972.

SUBHUMAN PRIMATES AS RESEARCH ANIMALS

Animals used for research are generally chosen for convenience rather than for rational, scientific reasons. This generalization applies whether large or small sized species are considered. Incidence of disease, handling requirements, supply, and cost are among factors considered when a species is chosen. A partial summary of animals used for research during 1971 is listed in Table 1. The data were obtained from a survey conducted by the Institute of Laboratory Animal Resources (NRC, 1972). As might be anticipated, rats and mice accounted for over 80% of the nearly 50 million animals used. Subhuman primates ranked 11th on the list and accounted for barely over 0.1% of the total. Typically, research with subhuman primates is done in large facilities with animal populations of over 300 (Table 2). Large facilities can efficiently and effectively provide quarantine, maintenance and husbandry services that are essential to prevent unit animal costs from becoming prohibitive. Currently wild-caught rhesus monkeys (♀, 4 kg) cost about $450 including quarantine, which represents a 3-fold increase during the past 6 years. Per diem charges in large facilities have remained about constant or even declined during the same period (NRC, 1975). The subject of usage and availability of subhuman primates for research has been recently reviewed (NRC, 1975).

TABLE 2

Primates used by size class of research facility[1]

| No. Primates | Facilities | | Primates Used | |
|---|---|---|---|---|
| | Total | Per Cent | Total | Per Cent |
| 1-99 | 189 | 69 | 6,776 | 13 |
| 100-299 | 50 | 18 | 8,381 | 15 |
| More than 300 | 37 | 13 | 39,900 | 72 |
| TOTAL | 276 | 100 | 55,057 | 100 |

[1]National Research Council, 1975

## USE AND CLASSIFICATION OF PRIMATES

Anatomical and physiological similarities that humans share with other primates contribute to the usefulness of subhuman species in biomedical research. Quantitative breakdown of particular types of research which utilize primates is given in Table 3. Pharmacology and toxicology, studies of disease (infectious, neoplastic, and organ system), and experimental surgery utilize over 60% of the research animals (NRC, 1975). The primary uses in pharmacology and toxicology are in preclinical safety evaluations. The application of metabolism data to the evaluation of drugs is the subject of a review (Drug Research Board, 1969).

TABLE 3

Biomedical research demand for primates, 1973[1]

| Research | Demand | |
|---|---|---|
| | Total | Per Cent |
| Pharmacology and toxicology | 10,499 | 25 |
| Vaccine production and safety testing | 4,944 | 12 |
| Diseases and experimental surgery | 14,654 | 36 |
| Neurophysiology | 6,599 | 16 |
| Physiology, reproduction, social behavior | 4,415 | 11 |
| TOTAL | 41,111 | 100 |

[1]National Research Council, 1975.

TABLE 4

Classification of living primates[1]

| Suborder | Families | |
|---|---|---|
| Prosimii | Tupaiidae<br>Lemuridae<br>Indriidae<br>Daubentoniidae<br>Lorisidae<br>Tarsiidae | Malagasy families |
| Anthropoidea | Callitrichidae<br>Cebidae<br>Cercopithecidae<br>Hylobatidae<br>Pongidae<br>Hominidae | New World Monkeys<br>Old World Monkeys<br>Apes and Humans |

[1] Napier and Napier, 1967.

From the 12 families of living primates (Napier and Napier, 1967), the number of species used in research is small. A classification of living primates (Table 4) includes the suborder Anthropoidea which can be divided into Old World (Africa and Asia) and New World (South and Central America) monkeys, apes and humans. Old and New World species are distinguished by the disposition of their nostrils, dentition, and structure of the skull. The rhesus monkey *Macaca mulatta*, an Old World species, is the primate of choice in about 60% of pharmacology and toxicology studies (NRC, 1975). Recently restricted supplies have resulted in increased uses of other species (Table 5). This listing of species imported to the United States is essentially a listing of the primary species used in research, since over 70% of imports are used within one calendar year. Trends in use patterns include increased use of New World monkeys, i.e., squirrel, night, woolly, and spider monkeys, capuchin and mystax marmosets, and a more diversified use of macaques (including stump-, pig-, and long-tailed) and vervet monkey. Wild-caught animals have historically dominated the primate trade, but restrictions on exports due in part to the uncertain status of wild populations have contributed to the development of a national plan to stabilize supply, including establishment of self-sustaining, domestic breeding colonies (Bermant and Lindburg, 1975; NRC 1975).

TABLE 5

Species of primates imported into the United States, 1970-72[1]

| Species | Number | Per Cent of Total |
| --- | --- | --- |
| Squirrel monkey | 81,296 | 34.4 |
| Rhesus monkey | 68,609 | 29.0 |
| Capuchin | 17,617 | 7.5 |
| Marmosets | 15,067 | 6.4 |
| Night monkey | 11,470 | 4.9 |
| Vervet | 9,195 | 3.9 |
| Woolly monkey | 6,595 | 2.8 |
| Spider monkey | 5,328 | 2.3 |
| Stump-tailed macaque | 3,953 | 1.7 |
| Long-tailed macaque | 4,733 | 2.0 |
| Baboon | 3,173 | 1.4 |
| Pig-tailed macaque | 1,679 | 0.7 |
| Chimpanzee | 624 | 0.3 |
| Other | 6,353 | 2.7 |
| TOTAL | 235,692 | 100.0 |

[1]National Research Council, 1975.

## COMPARATIVE METABOLISM IN LIVING THINGS

Metabolism is centrally important among the disposition processes, i.e. absorption, distribution, metabolism, and excretion, which determine the extent and duration of action of foreign compounds. Comparative metabolism studies form an important part of the scientific evaluation of the hazardousness of foreign compounds classed by use as drugs, food chemicals, pesticides, natural products and commercial chemicals (process, product, and pollutant substances). In predictive toxicology, extrapolation of results of comparative metabolism studies to man is a common goal. Knowledge of metabolic pathways and correlation of toxic effects with blood levels of parent compound and/or metabolites provide information that can contribute to establishment of safe patterns of use. During pesticide registration, metabolic data from a rodent and sometimes a nonrodent species are obtained. Such a narrow biological data base may not be sufficient to allow a solid prediction of the fate of an insecticide in persons engaged in manufacture, formulation, and application of the material or in those who are accidentally or adventitiously exposed. The Report of the Secretary's Commission (1969) called for improved techniques for prediction of human effects, "Improved scientific methods and protocols should be developed to assess dose-response and metabolic phenomena related to the biological effects of pest control chemicals in various species in order to increase the accuracy of extrapolative predictions concerning human effects." The same need was described by the Committee on Problems of Drug Safety (Drug Research Board, 1969): "Extended knowledge of pathways of drug metabolism in man and comparison with similar transformations in experimental animals is required for better prediction of likely alterations in drug metabolism in patients." It is likely that comparative metabolic studies utilizing primates will contribute to fulfillment of the needs expressed in these two statements.

Comparative metabolic studies may be considered as either chemical or biological in focus. In the former, the metabolic fates of a series of *foreign compounds* are studied in a particular test organism, most commonly the rat. Studies with a biological focus evaluate the metabolism of a given compound in a large number of *species*. Each approach has been productive, and they will be considered separately with emphasis being given to the biological approach.

*Chemical Focus*

Biotransformations of foreign compounds are apparently similar among living things. The biphasic nature of foreign compound metabolism has been emphasized by Williams (1974, 1976). First phase biotransformation involves oxidation, reduction, or hydrolysis, and in the second phase the products of the first phase are conjugated with natural body constituents to form polar excretory products. Within this basic scheme, considerable quantitative and qualitative variability is associated with both genetic and environmental factors.

Extensive studies, using primarily rodents, have clearly demonstrated that the enzymes catalyzing these reactions are generally *substituent specific* rather than *stereospecific* in the usual lock-key sense. Thus, for example, we do not hesitate to predict first phase formation of sulfoxides and esters, and amines from nitro compounds and second phase formation of conjugates with glucose, sulfate, amino acids and glutathione.

*Biological Focus*

*Comprehensive in vivo investigations*. Comparative studies of the fate of a particular foreign compound in a large series of living animals have been conducted less frequently than *in vitro* investigations. Results of *in vivo*

TABLE 6

Some foreign compounds metabolized similarly in vivo by humans and subhuman primates

| Compound | Primates ||| Metabolic Similarity ||
| | Human and Apes | Old World Monkeys | New World Monkeys | Oxidation, Reduction Hydrolysis | Conjugation |
|---|---|---|---|---|---|
| Arylacetic acids[1] | + | + | | | + |
| Benzoic acid[2] | + | + | + | | + |
| n-Butyl 4-hydroxy-3,5-diiodobenzoate[3] | + | + | | + | |
| Coumarin[4] | + | + | | + | |
| Ethanol[5] | + | + | | + | |
| Methanol[6] | + | + | | + | |
| Phenol[7] | + | + | | | + |
| Quinic acid[8] | + | + | | + | + |
| Sulphadimethoxine[9] | + | + | | | + |

[1]Williams, 1974; [2]Bridges et al., 1970; [3]Wold et al., 1973; [4]Gangolli et al., 1974; [5]Lieber and DeCarli, 1974; [6]Pieper and Skeen, 1973; [7]Capel et al., 1972; [8]Adamson et al., 1970a; [9]Adamson et al., 1970b.

investigations of this sort provide particularly valuable, basic knowledge about patterns of metabolism and insight into the relationship of these patterns to phylogeny. From the results of such studies has emerged the generalization that subhuman primates, especially Old World monkeys, frequently resemble humans in pathways of foreign compound metabolism (Williams, 1974, 1976). Some studies which support this statement are cited in Table 6. These examples, including references therein, represent comparative studies *in vivo* in which large numbers of species were used.

Two examples of comprehensive *in vivo* studies will be considered. The 7-hydroxylation of coumarin typifies common first phase metabolism in humans and Old World monkeys. Coumarin is a natural constituent of plants including tonka beans, dates and a variety of fruits. Hepatotoxicity resulted in termination of its use as a food additive in the United States. There are considerable differences in species sensitivity to coumarin, and therefore comparative metabolic studies were conducted (Table 7; Kaighen and Williams, 1961; Shilling et al., 1969; Gangolli et al., 1974). In humans the primary urinary metabolite is 7-hydroxycoumarin (about 80% of administered dose) and lesser amounts of *o*-hydroxyphenylacetic acid (6% or less) are also found (Shilling et al., 1969). In the rat and rabbit the latter is a predominant metabolite (12-27%), and it is coupled with large amounts (18-23%) of 3-hydroxycoumarin in the rabbit. Since this metabolite spectrum contrasts sharply with that of man, further comparative work was done by Gangolli et al. (1974). Using urines of 9 species dosed orally with coumarin they found considerable variability in the amount of 7-hydroxycoumarin excreted. In the squirrel monkey and 5 other species the amount of the 7-hydroxy derivative was 5% or less (Table 8). In sharp contrast, the baboon excreted 60% of the administered dose as 7-hydroxycoumarin, and hence this Old World species was the best metabolic model for humans.

A second comprehensive study is that of the metabolism of *n*-butyl 4-hydroxy-3,5-diiodobenzoate, an antithyroid drug. A study of its fate in 7 species including 5 primates revealed similarity of both first and second phase reactions in humans, rhesus monkeys and cynomologus monkeys (Wold et al., 1973). In humans, rhesus monkeys, rabbits and rats about 80% of the radioactivity was excreted in urine within 3 days. Fecal excretion was important in the New World species. Thin layer chromatographic analysis revealed 4-hydroxy-3,5-diiodohippurate and 4-hydroxy-3,5-diiodobenzoate in urine extracts of each species. Present exclusively in the urine extracts of the primates was 3,5-diiodo-4-methoxybenzoate which accounted for 3.6-21% of the urinary radioactivity. Metabolic similarity between humans and the subhuman primates, particularly

TABLE 7

Fate of coumarin in rats, rabbits and humans

| Species | % Dose Administered | |
|---|---|---|
| | Urine | Feces |
| Rat[1] | 47-61 | 33-38 |
| Rabbit[1] | 80-92 | 0.2-0.7 |
| Human[2] | 90-97 | --- |

References below Table 8.

TABLE 8

Urinary excretion of 7-hydroxycoumarin

| Species | % Administered Dose |
|---|---|
| Human[1] | 68-92 |
| Baboon[2] | 60 |
| Squirrel monkey[2] | 1 |
| Guinea pig[2] | 1 |
| Cat[2] | 19 |
| Dog[2] | 3 |
| Pig[2] | 12 |
| Ferret[2] | 1 |
| Rat[3] | 0.3-0.5 |
| Mouse[3] | 3 |
| Hamster[2] | 5 |
| Rabbit[3] | 10-16 |

[1]Shilling et al., 1969.
[2]Gangolli et al., 1974.
[3]Kaighen and Williams, 1961.

the Old World primates (rhesus and cynomologus monkeys) was established. Both New World monkeys (squirrel and capuchin) excreted up to 30 per cent of the administered dose in feces so this excretory pattern contrasted with that of the other species.

Comprehensive *in vivo* studies of insecticide metabolism in primates, analogous to the studies listed in Table 6, are lacking in the literature.

*In vitro approaches.* Invertebrates, lower vertebrates, and primates have been used in the study of comparative metabolism of foreign compounds including insecticides. However, use of primate tissues has been infrequent. Those methods listed here are ones likely to be productive in future comparative investigations. Killing of primates simply to obtain tissue to be used as an enzyme source is categorically inappropriate. Organ maintenance, cell culture and cell-free preparations which can make use of biopsy material must be further developed and utilized in comparative metabolism.

A fundamental difficulty in comparative studies is lack of information about fate of the insecticide in humans. Such knowledge would facilitate planning of chronic studies with animals and might be important in later epidemiological investigations. Sullivan and co-workers (Sullivan et al., 1972; Chin et al., 1974) have surmounted the problem by use of organ maintenance methods and chromatographic identification of metabolites. The procedures have considerable promise for future comparative work.

Carbaryl was selected as the prototype because of extensive metabolic studies that have been made in man and other species (Knaak et al., 1968; Sullivan et al., 1972). Surgical biopsies (500 mg) of rat, dog, guinea pig and human liver (Sullivan et al., 1972) and recently human tissues including liver, lung, kidney and placenta (Chin et al., 1974) were incubated in oxygenated media for 18 hours with carbaryl. Ethanolic extracts were analyzed by liquid scintillation counting (LSC) and DEAE-cellulose column chromatography.

Results obtained *in vitro* using napthyl-$^{14}$C-carbaryl in a human liver biopsy preparation obtained from an adult female (Sullivan et al., 1972) were qualitatively similar to *in vivo* results obtained by fluoresence analysis of DEAE-cellulose column chromatography fractions of urine of a male volunteer (Fig. 1). Important peak identifications include: D, 5,6-dihydro-5,6-dihydroxycarbaryl glucuronide + unknown glucuronides; F, 4-hydroxycarbaryl glucuronide; G, naphthyl glucuronide; I, 4-hydroxycarbaryl sulfate; J, naphthyl sulfate. The main difference between the 2 profiles of Fig. 1 is the greater naphthyl sulfate (J) *in vivo* and greater naphthyl glucuronide (G) *in vitro*. The difference may result from depletion of the sulfate pool during the tissue incubation period as is known to occur *in vivo* (Williams, 1959). The effects of substrate level need further evaluation to help define incubation conditions which will best reflect *in vivo* metabolic capabilities. The procedures provide a direct means of determination of metabolic profiles in humans or other animals which cannot be directly exposed.

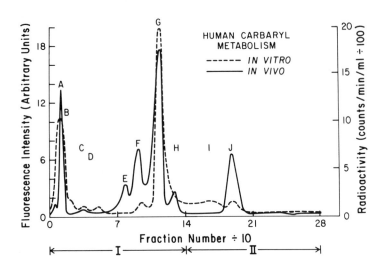

Fig. 1. DEAE-cellulose chromatograms of in vivo (fluorescence) and in vitro ($^{14}C$-naphthyl) carbaryl metabolites from human urine and liver biopsy preparations. Gradient elution: (I) 0.01-0.05, (II) 0.05-0.1 M Tris·HCl buffer, pH 7.5. Redrawn after Sullivan et al. (1972).

Mammalian cell cultures can provide useful test systems for comparative studies. Of particular interest for metabolism work are isolated hepatocyte cultures. Nebert and Gelboin (1969) first reported use of cultured explants or dispersions of fetal mouse liver tissue for drug metabolism studies. More recently, successful preparation of hepatocytes of adult rats has enabled study of kinetics of oxidation of ethylmorphine (Erickson and Holtzman, 1976) and alprenolol (Moldeus et al., 1974). Cultured human liver cells have been used to study the uptake and persistence of DDT, aldrin, dieldrin, parathion and chlordimeform (Murakami and Fukami, 1976), and rhesus monkey liver preparations have been used to investigate nicotine metabolism (Poole and Urwin, 1976). Cell culture technology will likely play a role in future comparative studies of the disposition of insecticides and other chemicals in primates.

Tissue homogenates are the most frequently employed enzyme source used in comparative metabolic studies. These cell-free systems may be easily manipulated, and their use has provided a wealth of biochemical information. Tissue availability severely limits primate studies since autopsy specimens are the usual enzyme sources.

The enzyme and biochemical composition of smooth and rough microsomal membranes from livers of rhesus monkeys were studied (Gram et al., 1971). Activities of 3 oxidative enzymes and UDP-glucuronyl transferase were about 2-fold greater in smooth than in rough membrane fractions. A comparative study published by Litterst et al. (1976) reported 14 metabolic parameters of hepatic microsomal and soluble fractions obtained from adult female and male rhesus monkeys, squirrel monkeys, Hanford miniature pigs, tree shrews *Tupaia glis*, and Sprague-Dawley rats. The animals were killed, livers removed and homogenates prepared by standard methods. Microsomal fractions were sedimented from the postmitochondrial supernatant, and enzyme assays were performed on the resulting fractions without delay. There was consistently a similar level of monooxygenase activity between the sexes except for the well established differences in the rat liver preparations (Table 9). Soluble transferase activity was likewise similar in the 5 species tested. Large variability between individuals observed by Litterst et al. (1976) was ascribed to use of wild-caught animals. Whether the variability was genetic or environmental might have been revealed by assay of a series of surgical biopsies taken over an extended period. Variability in liver enzyme activity of recently imported animals is not surprising. Regardless, the results of this study are useful as the most comprehensive study of microsomal enzymes using primates, and it will furnish useful baseline data for future *in vitro* investigations.

*Insecticides and Enzyme Induction*

With recognition of the importance of the liver in the metabolism of foreign compounds, concern has increased about chemicals which can alter the metabolic capability of liver enzyme systems, particularly NADPH-dependent monooxygenases. Enzyme induction is one of the important responses of living things to certain chemical stressors including some insecticides (see Conney and Burns, 1972). These adaptive responses may be ubiquitous in animals, but they are best characterized in rats. Some of the responses of animals, particularly primates, to liver enzyme inducers will be briefly described. Review of these studies will identify some experimental end points that have been utilized in our current studies in rhesus monkeys.

Hart et al. (1963) associated spraying of chlordane in animal quarters with shortened hexobarbital sleep times, liver hypertrophy, and increased liver microsomal enzyme activity in rats. Shortly thereafter the phenomenon of liver enzyme induction was studied in squirrel monkeys (Cram et al., 1965).

TABLE 9

*Microsomal NADPH monooxygenase activity of livers of 5 species[1]*

| Species | Sex | Cytochrome P-450[2] | N-Demethylation[3] | | Hydroxylation[3] | |
|---|---|---|---|---|---|---|
| | | | Ethylmorphine | Aminopyrine | Aniline | Biphenyl |
| Rhesus Monkey | M | 0.12 | 7.4 | 9.2 | 0.7 | 0.7 |
| | F | 0.12 | 7.2 | 9.9 | 0.6 | 0.6 |
| Rhesus Monkey | M | 0.12 | 7.4 | 9.2 | 0.7 | 0.7 |
| | F | 0.12 | 7.2 | 9.9 | 0.6 | 0.6 |
| Squirrel Monkey | M | 0.04 | 4.6 | 3.9 | 0.3 | 0.3 |
| | F | 0.04 | 4.4 | 3.3 | 0.3 | 0.3 |
| Tree Shrew | M | 0.05 | 6.7 | 5.4 | 0.2 | 1.1 |
| | F | 0.06 | 6.7 | 6.3 | 0.2 | 1.8 |
| Pig | M | 0.09 | 6.7 | 6.5 | 0.3 | 1.0 |
| | F | 0.09 | 6.7 | 6.9 | 0.5 | 1.1 |
| Rat | M | 0.10 | 8.6 | 7.1 | 0.4 | 0.7 |
| | F | 0.09 | 2.6 | 4.5 | 0.4 | 0.6 |

[1]Litterst et al., 1976.

[2]nmole P-450/mg protein.

[3]nmole product/mg protein/minute.

Gamma-Chlordane (10 mg/kg) was administered ip for 7 days. On the 8th day enzyme activities and cytochrome P-450 levels were measured in the 9000 g x 1 h supernatant of liver homogenates. Hexobarbital, 3,4-benzpyrene, aminopyrine, and zoxazolamine oxidations and $p$-nitrobenzoic acid reduction were increased in the chlordane-treated monkeys. Most dramatic was the nearly 60-fold increase in aminopyrine $N$-demethylation. Although the animals maintained their weights during the test period, fresh liver weights and total nitrogen in liver supernatant were 1.32- and 1.33-fold greater than controls. Liver hypertrophy and proliferation of endoplasmic reticulum were linked with weight and nitrogen changes observed. Five and one-half fold increases in cytochrome P-450 and 3-fold increases in NADPH oxidase activity were also observed. These responses of squirrel monkeys to chlordane treatment are qualitatively similar to those of rats observed earlier in the same laboratory (Hart et al., 1963).

In a second primate study, adult, male squirrel monkeys were dosed ip with either 5 or 10 mg DDT/kg/day for 7 days (Juchau et al., 1966). The monkeys were killed a day later and liver preparations biochemically assayed. Both DDT treatments markedly enhanced ring oxidation of zoxazolamine, $p$-nitrobenzoic acid reduction, aminopyrine $N$-dealkylation and increased to lesser extents 3,4-benzpyrene hydroxylation and hexobarbital side chain oxidation. The activities at the 2 dosages were not significantly different perhaps indicating maximal induction. DDT treatment also resulted in greater than 2-fold increases in microsomal cytochrome P-450 and NADPH oxidase activity.

Chadwick et al. (1971) measured D-glucuronic/D-glucaric acid ratios and liver enzyme activities in squirrel monkeys induced by DDT (5 mg daily, po) and ascorbic acid deficiency. In a more recent study squirrel monkeys were given DDT in peanut oil by stomach tube at 0.05, 0.5, 5 or 50 mg/kg/day (Cranmer et al., 1972). All animals at the highest dose died within 14 weeks. Increased ethyl $p$-nitrophenyl phenylphosphonothioate hydrolysis was obtained after 2 months at 5 mg DDT/kg/day. Also enhanced $p$-nitroanisole $O$-demethylation activity was found after 2 months at 0.5 and 5 mg DDT/kg/day. The study was continued 6 months but enzyme activities were not further increased.

Increased levels of cytochrome P-450, the terminal oxidase in the microsomal electron transport system, have been related to enzyme induction (see Conney and Burns, 1972). Following 69 to 74 months of treatment with dieldrin (0.01-5.0 ppm in diet), the livers of 6-year-old rhesus monkeys had higher cytochrome P-450 contents than unexposed controls (Wright et al., 1972). The mean level in preparations of controls was 1.20 nmole cytochrome P-450/mg microsomal protein, and 1.41, 1.43, 1.57, 1.69, 1.86 and 2.80 nmole/mg were obtained at dietary levels of 0.01, 0.1, 0.5, 1.0, 1.75, and 5.0 ppm dieldrin, respectively. Only at the 2 higher dieldrin levels were increased rates of $O$-demethylation of chlorfenvinfos obtained and no increases in either $p$-nitroanisole $O$-demethylation or aniline hydroxylation were measured. These observations show the usefulness of a battery of parameters in comparative studies of microsomal enzyme systems since the systems are differentially sensitive to inducers.

These studies with subhuman primates demonstrate the responsiveness of subhuman primates to chlorinated hydrocarbon liver enzyme inducers. The major objective of studies which will now be described is development of sensitive methods for continual assessment of liver oxidase activity in rhesus monkeys.

EXPERIMENTAL: CONTINUAL ASSESSMENT OF LIVER OXIDASE ACTIVITY IN RHESUS MONKEYS FED DDT

During the past 4 years, we have been developing methods to continually assess the liver oxidative metabolic capability of rhesus monkeys given low levels of DDT (10-500 ppm) in the diet. Plasma half-lives of antipyrine, levels of normal body constituents, and biochemical studies employing biopsy specimens have been used. Distinctive features of the study include use of animals as their own controls and routine reuse of the animals following completion of particular aspects of the work. Three of the 6 animals presently being used have been used for comparative metabolic studies for 6 years and foreign compounds to which they have been experimentally exposed include ethanol, phenobarbital and DDT. They have presented no unusual disease and their routine clinical chemistry is unremarkable.

*Animals and Chemical Treatments*

*Monkeys.* Male rhesus monkeys are quartered at the California Primate Research Center on the campus of the University of California, Davis. A group of three 8-11 kg monkeys and a group of three 4-6 kg monkeys are used. They are housed individually in stainless steel squeeze cages (LabCare[R], Research Equipment Co., Bryan, TX) to facilitate handling. A constant temperature 25 ± 4°C and 14L/10D photoperiod were maintained. The monkeys were fed Purina Monkey Chow-15 (Ralston Purina, St. Louis, MO). Water was provided *ad libitum*.

The animals receive regular veterinary care and all drug administrations are carefully recorded.

*Anesthesia*. Ketamine (12 mg/kg, im) a dissociative anesthetic with rapid onset and short duration of action (Chang and Glazko, 1974), was used for chemical restraint of monkeys during handling and biopsy procedures. In studies to be reported elsewhere, we have evaluated the potency of ketamine as an inhibitor and inducer of liver monooxygenase activity. These studies included measurement of liver levels of ketamine and its primary metabolites as well as liver enzyme assays. We have concluded that ketamine can interfere with measurement of the oxidative metabolic capability of rhesus monkeys; but, under the use conditions employed in our routine operations, the experimental parameters are not measurably affected. Potential interactions of this sort are particularly important in long term studies.

*Administration of DDT*. DDT (99+ % pure; Aldrich Chem. Co., Milwaukee, WI) was administered by applying acetone solutions (100 µl) to pellets of monkey chow so the final concentration of DDT was 10, 100, or 500 ppm. Assuming consumption of 250 g food by a 5 kg monkey, a dietary level of 10 ppm DDT would result in a DDT dosage of 0.5 mg/kg/day. No chlorohydrocarbons were detectable (<10 ppb) in untreated food, and each lot of treated food was analyzed to confirm the actual level in the diet. In addition to being a common route of human exposure, the feeding of treated diet is less traumatic than either stomach intubation or parenteral injection. Although monkeys are frequently suspicious of unusual odors and flavors in food and water, the levels of DDT used in these studies had no deleterious effects. During the exposure period, the monkey weights were stable and food consumption normal.

*DDT metabolites in blood and fat*. Blood (Palmer et al., 1972) and fat (Barquet et al., 1972) of monkeys fed 0, 10, 100 and 500 ppm DDT in diet was analyzed for DDT and metabolites. Dose-dependent absorption is clearly indicated (Fig. 2) by the progressive increases in DDT in blood for the 3 levels of DDT in diet. The periods of exposure were determined by intervals required for measurement of the several experimental parameters. Small amounts of DDD and DDE were detected in blood extracts, but they could not be realiably measured on a routine basis. Quantitative determination of DDT, DDE, and DDD in subcutaneous abdominal fat was made (Fig. 3). DDT and each metabolite increased with dose consistent with product precursor relationships. These results are more similar to those obtained in humans (Morgan and Roan, 1971) than to previous results obtained in rhesus monkeys (Durham et al., 1963). Durham et al. concluded the monkey incapable of formation of DDE. Although the nature of the system promoting

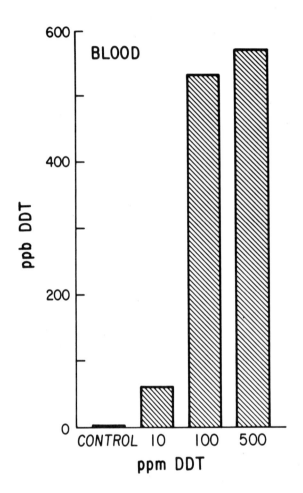

Fig. 2. Blood levels of DDT in a male, adult rhesus monkey before and following successive 30-day feedings of diet containing 10, 100, and 500 ppm DDT. Hexane extracts analyzed by electron-capture gas-liquid chromatography.

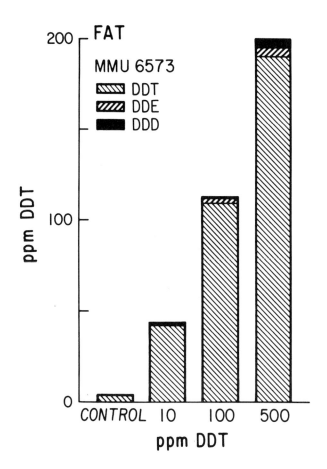

Fig. 3. Levels of DDT, DDE, and DDD in abdominal subcutaneous fat of a male rhesus monkey (ca. 6 kg) following ingestion for successive 30-day periods of diet containing 10, 100, and 500 ppm DDT.

TABLE 10

Plasma antipyrine half-lives in humans and other species

| Species | Antipyrine Half-life (hours) | Reference |
|---|---|---|
| Human | 12.0 | Brodie et al., 1958 |
|  | 13.1 | Kolmodin et al., 1969 |
|  | 12.4 | Vessel et al., 1969 |
| Rhesus Monkey | 1.8 | Burns, 1966 |
|  | 1.2 | Hucker et al., 1972 |
|  | 1.7 | Branch et al., 1974 |
|  | 1.3 | This investigation |
| Dog | 1.7 | Burns, 1966 |
|  | 2.5 | Vessell et al., 1973 |
| Rat | 2.4 | Brodie et al., 1958 |
| Rabbit | 1.1 | Brodie et al., 1958 |
|  | 1.1 | Statland et al., 1973 |
| Mouse | 0.2 | Brodie et al., 1958 |

DDT dehydrochlorination has not been established, there is no doubt that DDE is a DDT metabolite in rhesus monkeys. The ratio of DDT:DDE:DDD at the end of the 100 ppm DDT feeding was approximately 225:6:1 and after 500 ppm 220:13:10. The metabolic changes responsible for the dramatic relative increases in both DDE and DDD are uncertain and will be more thoroughly evaluated in later studies. Since DDE is a prominent DDT metabolite and inducer of liver oxidases in rats (Fishbein, 1974; Peterson and Robinson, 1964), rhesus monkeys may be a better human model than rats for liver oxidase induction studies.

*Administration of phenobarbital.* Phenobarbital (15 mg/kg) was administered im twice daily for 4 days to 2 of the 3 monkeys at the end of the DDT treatment period.

*In Vivo Indicators of Liver Function*

*Plasma half-life of antipyrine.* Antipyrine (2,4-dimethyl-1-phenyl-3-pyrazolin-5-one) is useful in comparative metabolic studies as an indicator of liver monooxygenase activity. In humans antipyrine is readily absorbed, not excreted unchanged, is metabolized in liver and negligibly bound to plasma protein (Brodie et al., 1949). Antipyrine plasma half-lives have been measured in several species including humans and rhesus monkeys (Table 10). Vesell and Page (1968) in studies with identical and fraternal twins demonstrated that antipyrine half-lives in humans are primarily genetically determined thus supporting the use of protocols in which animals are used as their own controls.

There is considerable variability among rates of elimination of antipyrine in different species. The plasma half-lives range between 0.2 hours in the mouse to about 12 hours in humans. The rhesus monkey could be classed as intermediate since the antipyrine plasma half-life is 1.2-1.8 hours (Table 10). In spite of species differences the parameter is a useful one in comparative studies. Animals can be used as their own controls and can be frequently retested.

The plasma half-life of antipyrine in a group of Swedish workers occupationally exposed to DDT and lindane was lower than the half-life of an unexposed control group (Kolmodin et al., 1969). Since measurements of antipyrine plasma levels can be simply made using spectrophotometry (Brodie et al., 1949), and since the chemical is of low toxicity, minimal risk is associated with *in vivo* antipyrine studies.

We have used $^{14}$C-antipyrine and LSC analysis for measurement of plasma half-lives in rhesus monkeys. The animals are held in restraining chairs (Reubner et al., 1975) and the experimental period is initiated by iv administration (saphenous vein) of 15 mg $^{14}$CH$_3$-$N$-antipyrine (0.25-0.5 µCi)/kg in sterile saline. At 30, 60, 90 and 120 minutes about 3 ml blood are drawn from the brachial vein, plasma and erythrocytes are separated by centrifugation, and 1 ml aliquots of plasma are transferred to a tube containing 3 ml 0.33 N NaOH. Extraction with 20 ml 12% isoamyl alcohol in toluene (v/v) removes unmetabolized antipyrine. Fifteen ml are transferred to a vial containing a concentrated scintillation fluid (0.6 ml solution which contains 0.125 g $p$-bis 2-(5-phenyloxazolyl)-benzene + 10 g 2,5-diphenyloxazole in 100 ml toluene) and antipyrine is determined by LSC. Plasma antipyrine concentration is expressed as a logarithmic function of time and curves are determined by linear regression analysis.

In Fig. 4 are profiled antipyrine half-lives before, during and after feeding of the DDT-treated diets. Significant ($p < 0.05$) shortening of the half-life compared to that in the control period was obtained after 30 days of feeding 10 ppm DDT and through the remainder of the 60-day period during which the monkeys were fed 100 and 500 ppm DDT. At the end of the DDT exposures, two of the 3 monkeys were treated with phenobarbital (15 mg/kg x 2 x 4 days, im) to determine whether further enhancement of hepatic oxidase activity could be obtained. An additional 20% reduction in antipyrine plasma half-life was obtained in the phenobarbital treated monkeys.

Fig. 4 shows that the half-lives remained below pre-DDT control values for nearly 50 days. Earlier data of Remmer et al. (1968) obtained in dogs allowed clear distinction between short-term phenobarbital induction and long-term DDT induction. DDT blood levels were not reduced within 21 days by phenobarbital and the slow increases in antipyrine half-lives seem to be associated with DDT persistence. The kinetics of DDT elimination will be studied in detail in future investigations.

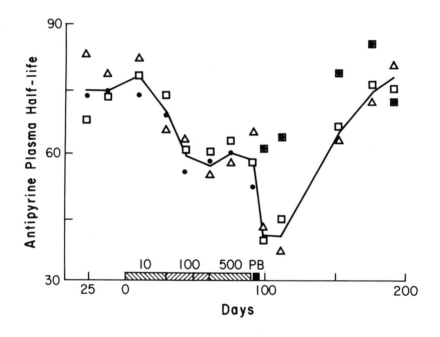

*Fig. 4. Antipyrine plasma half-lives in 3 male, rhesus monkeys before and during DDT feeding, after phenobarbital treatments, and during recovery period. Boxed solid circles indicate values for animal not treated with phenobarbital.*

*Normal body constituents.* Elevated activity of liver enzymes has been associated with increased urinary excretion of 6β-hydroxycortisol (Birchall et al., 1966; Bledsoe et al., 1964; Kuntzman et al., 1968; Poland et al., 1970) and D-glucaric acid (Hunter et al., 1971; Morgan and Roan, 1974; Sotaniemi and Medzihradsky, 1973). These factors may be particularly useful as indirect indicators of liver function in epidemiological studies.

6β-Hydroxycortisol excretion was increased in persons occupationally exposed to DDT (Poland et al., 1970) who were previously shown to have no liver disease or hepatotoxicity (Law et al., 1967). In addition, 6β-hydroxycortisol levels in urine were elevated in persons with occupational endrin exposures (Jager, 1970). Increased 6β-hydroxycortisol levels represent a biological response which reflects altered liver function in the absence of clinical dysfunction in the usual sense. This parameter is increased in rhesus monkeys treated with phenobarbital (unpublished observations) confirming the earlier observations of Birchall et al. (1966) using white-fronted capuchins *Cebus albifrons*. The usefulness of this parameter as a sensitive indicator of altered liver function is under current investigation.

D-glucaric acid excretion is a second natural urine constituent that has been used as an indicator of elevated liver enzyme activity following experimental drug treatments (Hunter et al., 1971) and occupational cyclodiene exposure (Hunter et al., 1972). D-glucaric acid is a product of glucuronic acid metabolism. Increases in diphosphoglucose dehydrogenase, glucuronolactone dehydrogenase, and glucurone δ-lactone hydrolase associated with liver enzyme induction result in elevated urinary D-glucaric acid levels in primates and the guinea pig. In rats ascorbic acid is the corresponding glucuronic acid metabolite. D-glucaric acid in urine may be sensitively analyzed by the Marsh procedure (Marsh, 1963). Phenobarbital treatment results in elevated D-glucaric acid excretion in rhesus monkeys (unpublished observations). Current DDT induction studies include measurement of D-glucaric acid excretion rates.

## In Vitro Indicators of Liver Function

*Closed needle biopsy*. Menghini (1958) aspiration techniques for liver biopsy have been used in humans for diagnostic purposes. Adaptations of those procedures for studies of liver ultra-structure and monooxygenase activity have been described (Krieger et al., 1976). The procedures permit continual monitoring of morphological and biochemical parameters in animals used as their own controls.

The basic biopsy needle assembly includes a skin trocar, cannula for cleaning, occluding nail, biopsy needle (1.9 x 70 mm) fitted with an adjustable guard to limit penetration, and disposable syringe (12 ml). Monkeys are deprived of food for 12 hours before biopsy. Sterile technique is employed, and ketamine (12 mg/kg, im) anesthesia is used. Compared to results obtained using the lateral recumbent position (Krieger et al., 1976), larger tissue yields (consistently about 80 mg) and less probing within the peritoneal cavity are possible by positioning the monkeys midway between the lateral recumbent and the prone positions. Full account of the procedures will be published (Miller et al., in manuscript).

During the past 4 years over 300 biopsies have been performed on rhesus monkeys with no associated mortality or morbidity. Some individuals have been biopsied over 30 times. Biopsy procedures are conservative practices which should be further developed, more widely used, and more completely evaluated in toxicology.

*Monooxygenase activities of liver homogenates.* Substrates for liver monooxygenase assays are characteristically lipophilic. In this study each was chosen for its characteristic pattern of microsomal metabolism. Rates of aldrin epoxidation, dihydroisodrin hydroxylation, *p*-nitroanisole *O*-demethylation, and 3,4-benzpyrene hydroxylation are routinely measured using homogenates of liver biopsy specimens as the enzyme source. Each substrate is stable and available in pure form, and each substrate (or product) is amenable to sensitive analysis.

The oxidations require NADPH and molecular oxygen, and incubation conditions are listed in Table 11. Experimental protocols minimize the time between biopsy and enzyme assay and also the amount of processing preceding analysis. Although data are not available to support the need for such procedures, they are considered particularly critical for low volume enzyme assay. Details of the assay and properties of the aldrin epoxidation and dihydroisodrin hydroxylation systems in monkey liver homogenates have been published (Krieger et al., 1976), and similar reports on other substrates will be made.

It has been experimentally established that amounts of product formed are proportional to homogenate protein and time and are independent of substrate level. Presented in Figs. 5, 6, and 7 are examples of the effect of protein, time, and substrate level on product formation. Dieldrin and monohydroxydihydroisodrin formation are proportional to homogenate protein up to about 2 mg/incubation. Protein levels of 0.5-0.8 mg/incubation are routinely used. The time course for 3,4-benzpyrene hydroxylation is shown in Fig. 6. Hydroxylation is usually measured employing a 30 minute incubation period. The substrate-product curve for *p*-nitroanisole *O*-demethylation is given in Fig. 7. Due to the low water solubility of most substrates used in microsomal oxidase studies, substrate-product relationships may be complex. Work on these biochemical parameters was done using biopsy-like liver cores rather than biopsy specimens *per se*.

TABLE 11

Low volume liver homogenate oxidate assay conditions

| Transformation | Substrate/Carrier | Time (min.) | Protein (mg/incub.) | Analysis |
|---|---|---|---|---|
| Aldrin epoxidation[1] | 20 µg/5 µl MeOH | 4 | 0.5-0.8 | Product/EC-GLC[2] |
| Dihydroisodrin hydroxylation[1] | 20 µg/5 µl MeOH | 4 | 0.5-0.8 | Product/EC-GLC[2] |
| p-Nitro-$^{14}CH_3$ anisole demethylation[1] | 120 µg/10 µl EtOH | 30 | 0.5-0.8 | Substrate/Partition, LSC[3] |
| 3,4-$^3$H-Benzpyrene hydroxylation[4] | 25 µg/50 µl Acetone | 15 | 0.1-0.2 | Substrate/$^3H_2O$, LSC[4] |

[1] Incubation mixture (0.2 ml) contains $KH_2PO_4$-$K_2HPO_4$ buffer (100 mM, pH 7.4), glucose-6-phosphate (2.3 mM), NADP (0.23 mM), glucose-6-phosphate dehydrogenase (1 unit/incub.) and KCl (2.7 mM). Temperature, 37°C.

[2] Krieger et al., 1976.

[3] Matthews and Casida, 1970.

[4] Hayakawa and Udenfriend, 1973. One ml incubation mixture contained $KH_2PO_4$-$K_2HPO_4$ (100 mM, pH 7.7), NADPH (0.4 mM), $MgCl_2$ (3 mM) and EDTA (0.1 mM).

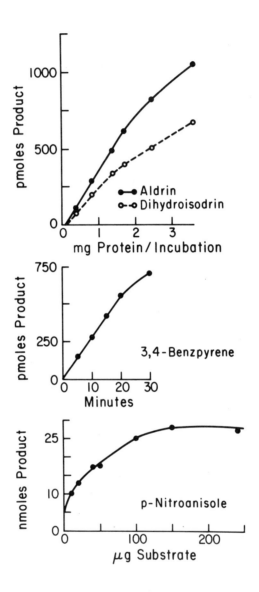

Fig. 5. (Top) Aldrin epoxidation and dihydroisodrin hydroxylation in 4 minute incubations at 37°C. Standard incubation media was used. Fig. 6. (Center) Benzpyrene oxidation over a 30 minute time course. Standard incubation media and 37°C were used. Fig. 7. (Bottom) Effect of substrate level on p-nitroanisole O-demethylation in standard media at 37°C.

If biopsy specimens are to be useful in assessment of monooxygenase capability, the specimen must be representative of the whole liver. Remmer et al. (1973) have cautioned that needle biopsy samples constitute only a fraction of the mass of the liver. This is emphasized by the fact that in a 5 kg rhesus monkey with liver mass of about 3% of total body mass (Bourne, 1975), a 75 mg biopsy sample constitutes only about 1/2000 of the liver.

TABLE 12

Monooxygenase activities of whole liver and corresponding needle biopsy specimens

| Substrate | Homogenate Activity[1] pmol/mg pro/min | | Correlation |
| --- | --- | --- | --- |
| | Needle | Whole | |
| Aldrin (n = 13) | 430 ± 313[2] | 361 ± 212 | 0.94 |
| Dihydroisodrin (n = 8) | 189 ± 123[2] | 150 ± 60 | 0.89 |
| p-Nitroanisole (n = 6) | 259 ± 93[2] | 227 ± 86 | 0.89 |
| 3,4-Benzpyrene (n = 4) | 18 ± 3 | 14 ± 1 | 0.46 |

[1] Livers obtained from male and female rhesus monkeys (1-10 kg). Final protein concentrations were 2-4 mg/ml. Volumes of "needle" incubations were 0.2 ml and homogenates were 2.0 ml.

[2] Significantly greater than whole liver homogenate activity using paired "T" test ($p < 0.05$).

Direct comparison of activities of homogenates of biopsy samples and whole liver homogenates is difficult. We have investigated the relationship by utilizing livers of animals killed for other purposes. From these livers, biopsy-like specimens were taken. The remaining liver was then homogenized and is listed as "whole homogenate" in Table 12. Homogenates of biopsy-like liver specimens had higher specific monooxygenase activities than whole liver homogenates. Highest activities were found for aldrin epoxidation in both types of preparations. In both cases oxidase activities ranked as follows: epoxidation > O-demethylation > hydroxylation > aryl hydroxylation. There was considerable variability in levels of activity measured in both types of preparations. Variability is in part due to use of animals of different ages, medical histories, methods of euthanasia, and uncertain environmental factors.

From these experiments we conclude that the biopsy specimens are useful enzyme sources for low volume determinations of liver monooxygenase activities.

Under optimized *in vitro* conditions (Table 11) biopsy specimens were assayed for aldrin epoxidation and *p*-nitroanisole *O*-demethylation activities during control, DDT-feeding, phenobarbital treatment, and post-treatment periods. Mean specific activities from 2 control period tests in which 3 monkeys were used were 200 pmol dieldrin/mg protein/min and 215 pmol *p*-nitrophenol/mg protein/min for epoxidation and *O*-demethylation activity, respectively. Following 30 days of feeding 10 ppm DDT, the oxidase levels were approximately doubled, and after successive 30-day periods at 100 and 500 ppm DDT the epoxidation and *O*-demethylation activities were increased 3- and 7.5-fold, respectively, relative to control activities. Increases in specific oxidase activity were also measured following the phenobarbital treatments. This indicates additional liver oxidase capacity even though antipyrine plasma half-lives (Fig. 4) and aldrin epoxidation activities had not increased markedly between feeding of 100 and 500 ppm DDT in diet. After 8 months when a second study was initiated monooxygenase activities were at control levels. The sensitivity of the rhesus monkeys to DDT induction is similar to that of rats as summarized by Fishbein (1974).

Biochemical monooxygenase assays clearly are feasible using closed needle biopsy specimens, and such assay procedures can provide information about the metabolic capability of liver enzyme systems without killing the test organisms.

CONCLUSIONS

The utility of rhesus monkeys as experimental animals in studies of liver enzyme induction has been demonstrated. Associated with dose-dependent increases in fat and blood levels of the inducer, decreased antipyrine plasma half-lives and increased rates of monooxygenase-catalyzed aldrin epoxidation and *p*-nitroanisole *O*-demethylation have been measured, indicative of induction. These changes were observed in the absence of any overt symptoms of toxicity or other clinical abnormalities. Hence, these changes represent experimental end-points that might be useful for prediction and study of effects of low level exposures.

Although only liver enzyme induction was studied, the general experimental approach involving the conservative use of subhuman primates may be employed for other studies of chemico-biological interactions.

It remains that subhuman primates are used too infrequently in studies of the toxicology of chemicals used for pest control.

Metabolic as well as morphological and physiological similarities between human and subhuman primates, particularly Old World monkeys, can be exploited in predictive toxicology. Particularly important are comprehensive comparative studies such as those of R. T. Williams (1974, 1976) in which the first and second phase metabolism of a foreign compound are investigated in a number of species. Results of such studies will form a data base which may make more rational the process of selection for species for toxicological research.

The precise role of subhuman primates in future biomedical research in toxicology is uncertain. General concern about and regulation of experimental studies with humans have increased (see National Academy of Science Forum, 1975). It seems likely that subhuman primates will become more important in the biological evaluation of drugs, food chemicals, pesticides and other commercial chemicals. However, limited primate resources will be an important factor influencing, and even limiting, patterns of the use in research.

Subhuman primates must be considered valuable allies rather than surrogates in comparative toxicological studies. Ethical and scientific concerns require that we undertake these studies with a profound respect for life.

ACKNOWLEDGEMENTS

Research support was obtained from NIH ES 00054 and ES 00125. JLM was a Shell Companies Foundation Fellow. Mr. Leonid Pulchritudoff and others associated with the California Primate Research Center have provided valuable assistance.

REFERENCES

Adamson, R. H., Bridges, J. W., Evans, M. E., Williams, R. T., Biochem. J. 116, 437 (1970a).

Adamson, R. H., Bridges, J. W., Kibby, M. R., Walker, S. R., Williams, R. T., Biochem. J. 118, 41 (1970b).

Barquet, A., Morgade, C., Cassady, J., "Reproducibility of Macro-Mills Procedure for Pesticide Residues in Fat," in Epidemiology of DDT, Davies, J. E., Edmunson, W. F., Eds., Futura Publ. Co., Inc., Mt. Kisco, N.Y., pp. 141-146 (1972).

Bermant, G., Lindburg, D. G., Eds., "Primate Utilization and Conservation," John Wiley and Sons, London, 196 pp. (1975).

Birchall, K., O'Day, W. G., Fajer, A. B., Burstein, S., Gen. Comp. Endocrinol. 7, 352 (1966).

Bledsoe, T., Island, D. P., Ney, R. L., Liddle, G. W., *J. Clin. Endocrinol. Metab.* 24, 1303 (1964).
Bourne, G. H., Ed., "The Rhesus Monkey," Vols. I and II., Academic Press, New York, N.Y. (1975).
Branch, R. A., Shand, D. G., Wilkinson, G. R., Nies, A. S., *J. Clin. Invest.* 53, 1101 (1974).
Bridges, J. W., French, M. R., Smith, R. L., Williams, R. T., *Biochem. J.* 118, 47 (1970).
Brodie, B. B., Axelrod, J., Soberman, R., Levy, B. B., *J. Biol. Chem.* 179, 25 (1949).
Burns, J. J., in Proc. 1966 Conf. Nonhuman Primate Toxicology, Miller, C. O., Ed., Govt. Printing Off., pp. 66-68 (1966).
Capel, I. D., French, M. R., Millburn, P., Smith, R. L., Williams, R. T., *Xenobiotica* 2, 25 (1972).
Chadwick, R. W., Cranmer, M. F., Peoples, A. J., *Toxicol. Appl. Pharmacol.* 20, 308 (1971).
Chang, T., Glazko, A. J., *Int. Anesthesiol. Clin.* 12, 157 (1974).
Chin, B. H., Eldridge, J. M., Sullivan, L. J., *Clin. Toxicol.* 7, 37 (1974).
Conney, A. H., Burns, J. J., *Science* 178, 576 (1972).
Coulston, F., "Qualitative and Quantitative Relationships Between Toxicity of Drugs in Man, Lower Mammals, and Nonhuman Primates," in Proc. 1966 Conf. Nonhuman Primate Toxicology, Miller, C. O., Ed., Govt. Printing Off., pp. 31-47 (1966).
Coulston, F., Serrone, D. M., *Ann. N.Y. Acad. Sci.* 169, 681 (1969).
Cranmer, M., Peoples, A., Chadwick, R., *Toxicol. Appl. Pharmacol.* 21, 98 (1972).
Cram, R. L., Juchau, M. R., Fouts, J. R., *J. Lab. Clin. Med.* 66, 906 (1965).
Drug Research Board, *Clin. Pharm. Ther.* 10, 607 (1969).
Durham, W. F., Ortega, P., Hayes, W. J., Jr., *Arch. Int. Pharmacodyn. Ther.* 141, 111 (1963).
Erickson, R. R., Holtzman, J. L., *Biochem. Pharmacol.* 25, 1501 (1976).
Fishbein, L., *J. Chromatog.* 98, 177 (1974).
Gangolli, S. D., Shilling, W. H., Grasso, P., Gaunt, I. F., *Biochem. Soc. Trans.* 2, 310 (1974).
Gram, T. E., Schroeder, D. H., Davis, D. C., Reagan, R. L., Guarino, A. M., *Biochem. Pharmacol.* 20, 1371 (1971).
Hart, L. G., Shultice, R. W., Fouts, J. R., *Toxicol. Appl. Pharmacol.* 5, 371 (1963).
Hayakawa, T., Udenfriend, S., *Anal. Biochem.* 51, 501 (1973).
Heywood, R., Med. Primatol. 1972. Proc. 3rd. Conf. Exp. Med. Surg. Primates, Lyon, pp. 95-99 (1972).

Hucker, H. B., Stauffer, S. C., White, S. E., *J. Pharmaceut. Sci.* 61, 1490 (1972).
Hunter, J., Maxwell, J. D., Stewart, D. A., Williams, R., Robinson, J., Richardson, A., *Nature* 237, 399 (1972).
Jager, K. W., Aldrin, Dieldrin, Endrin and Telodrin: An Epidemiological and Toxicological Study of Long Term Occupational Exposure, Elsevier, Amsterdam (1970).
Juchau, M. R., Gram, T. E., Fouts, J. R., *Gastroenterol.* 51, 213 (1966).
Kaighen, M., Williams, R. T., *J. Med. Pharm. Chem.* 3, 25 (1961).
Knaak, J. B., Tallant, M. J., Kozbelt, S. J., Sullivan, L. J., *J. Agric. Food Chem.* 16, 465 (1968).
Kolmodin, B., Azarnoff, D. L., Sjoqvist, F., *Clin. Pharm. Therap.* 10, 638 (1969).
Krieger, R. I., Miller, J. L., Gee, S. J., Thongsinthusak, T., *Drug Metab. Dispos.* 4, 28 (1976).
Kuntzman, R., Jacobson, M., Levin, W., Conney, A. H., *Clin. Pharmacol. Therap.* 11, 724 (1968).
Laws, E. R., Curley, A., Biros, F. J., *Arch. Environ. Health* 15, 766 (1967).
Lieber, C. S., DeCarli, L. M., *J. Med. Prim.* 3, 153 (1974).
Litterst, C. L., Gram, T. E., Mimnaugh, E. G., Leber, P., Emmerling, D., Freudenthal, R. I., *Drug Metab. Dispos.* 4, 203 (1976).
Marsh, C. A., *Biochem. J.* 86, 77 (1963).
Matthews, H. B., Casida, J. E., *Life Sci.* 9, 989 (1970).
Menghini, G., *Gastroenterol.* 35, 190 (1958).
Miller, C. O., Ed., Conference on Nonhuman Primate Toxicology, Washington, D.C., p. 168 (1968).
Moldeus, P., Grundin, R., Vadi, H., Orrenius, S., *Eur. J. Biochem.* 46, 351 (1974).
Morgan, D. P., Roan, C. C., *Arch. Environ. Health* 22, 301 (1971).
Morgan, D. P., Roan, C. C., "The Metabolism of DDT in Man," in Essays in Toxicology, Hayes, W. J., Jr., Ed., 5, 39 (1974).
Murakami, M., Fukami, J.,-I, *Bull. Environ. Contam. Toxicol.* 15, 425 (1976).
Napier, J. R., Napier, P. H., "A Handbook of Living Primates," Academic Press, New York, N.Y. (1967).
National Academy of Sciences Forum, "Experiments and Research with Humans: Values in Conflict," National Academy of Sciences, Washington, D.C., p. 234 (1975).
National Research Council, "Annual Survey of Animals Used for Research Purposes During Calendar Year 1971," Institute of Laboratory Animal Resources, Washington, D.C., p. 12 (1972).

National Research Council, "Nonhuman Primates: Usage and Availability for Biomedical Programs," Institute of Laboratory Animal Resources, National Academy of Science, Washington, D.C., p. 122 (1975).
Nebert, D. W., Gelboin, H. Y., in Microsomes and Drug Oxidations, Gillette, J. R., et al., Eds., Academic Press, New York, p. 389 (1969).
Palmer, L., Kolmodin, B., Hedman, B., J. Chromatog. 74, 21 (1972).
Peterson, J. E., Robinson, W. H., Toxicol. Appl. Pharmacol. 6, 321 (1964).
Pieper, W. A., Skeen, M. J., Biochem. Pharmacol. 22, 163 (1973).
Poland, A., Smith, D., Kuntzman, R., Jacobson, M., Conney, A. H., Clin. Pharm. Therap. 11, 724 (1970).
Poole, A., Urwin, C., Biochem. Pharmacol. 25, 281 (1976).
Remmer, H., Estabrook, R. W., Schenkman, J., Greim, H., "Induction of microsomal liver enzymes," in Enzymatic Oxidation of Toxicants, Hodgson, E., Ed., North Carolina State Univ., Raleigh, p. 65 (1968).
Remmer, H. B., Schoene, B., Fleishman, R., Held, H., "The Liver, Quantitative Aspects of Structure and Function," Baumgartner, G., Preisig, R., Eds., S. Karger, Basel, p. 232-239 (1973).
Ruebner, B. H., Krieger, R. I., Miller, J. L., Tsao, M., Rorvik, M., "Hepatic and Metabolic Effects of Ethanol on Rhesus Monkeys," in Experimental Studies of Alcohol Intoxication and Withdrawal, Gross, M. M., Ed., 59, 395 (1975).
Reulis, H. W., Drug Metab. Revs. 4, 115 (1973).
Shilling, W. H., Crampton, R. F., Longland, R. C., Nature (London) 221, 664 (1969).
Smith, C. C., "Role of nonhuman primates in predicting metabolic disposition of drugs in man," in Proc. 1966 Conf. Nonhuman Primate Toxicology, Miller, C.O., Ed., Govt. Printing Off., pp. 57-66 (1966).
Smith, C. C., Weigel, W. W., Wolfe, G. F., Mattingly, S., Med. Primatol. 1972, Proc. 3rd Conf. Exp. Med. Surg. Primates, Lyon, pp. 279-285 (1972).
Smith, R. L., Williams, R. T., J. Med. Primatol. 3, 138 (1974).
Sotaniemi, E. A., Medzihradsky, F., Clin. Pharmacol. Ther. 14, 147 (1973).
Statland, B. E., Astrup, P., Black, C. H., Oxholm, E., Pharmacol. 10, 329 (1973).
Sullivan, L. J., Chin, B. H., Carpenter, C. P., Toxicol. Appl. Pharmacol. 22, 161 (1972).
Udall, V., Proc. European Soc. Study Drug. Tox. 11, 57 (1969).

U. S. Department of Health, Education, and Welfare," Report of the Secretary's Commission on Pesticides and Their Relationship to Environmental Health," Govt. Printing Off., Washington, D.C., p. 677 (1969).

Vesell, E. S., Page, J. G., *Science* 161, 72 (1968).

Vesell, E. S., Lee, C. J., Passananti, G. T., Shively, C. A., *Pharmacol.* 10, 317 (1973).

Williams, R. T., "Detoxication Mechanisms," 2nd Ed., Chapman and Hall Ltd., London, (1959).

Williams, R. T., *Ann. N.Y. Acad. Sci.* 179, 141 (1971).

Williams, R. T., *Biochem. Soc. Trans.* 2, 359 (1974).

Williams, R. T., Personal communication (1976).

Wold, J. S., Smith, R. L., Williams, R. T., *Biochem. Pharmacol.* 22, 1865 (1973).

Wright, A. S., Potter, D., Wooder, M. F., Donniger, C., Greenland, R. D., *Fd. Cosmet. Toxicol.* 10, 311 (1972).

METABOLISM OF INSECT GROWTH REGULATORS IN ANIMALS

G. W. Ivie

*Veterinary Toxicology and Entomology
Research Laboratory, Agricultural Research
Service, U. S. Department of Agriculture*

ABSTRACT. *Compounds that selectively disrupt the normal growth and development processes of insects offer a new approach to insect control with minimal adverse environmental interactions. Studies with several insect juvenile hormone analogs have shown that these compounds are highly biodegradable in most animal species and that residues are generally rapidly excreted from the body. Results of preliminary studies in the author's laboratory with the chitin synthesis inhibitor, diflubenzuron, indicate that this compound is readily metabolized by cattle and sheep but much less so by the stable fly and housefly. Residues are not retained within tissues of the treated ruminants to any appreciable extent.*

INTRODUCTION

Attempts to exploit the biochemical differences between insects and most other animals in the development of selective insecticides have within the past decade come to fruition. With the isolation and structure elucidation of the major insect juvenile hormone that controls maturation in several insect species (Roller et al., 1967), the foundation was laid for the subsequent and highly intensive efforts toward developing synthetic analogs of these compounds for practical insect control. Another group of insect growth regulators, the chitin synthesis inhibitors (van Daalen et al., 1972; Wellinga et al., 1973a,b) also offer excellent potential for controlling numerous insect pests with apparently little or no adverse interactions with most nontarget species.

Although the insect growth regulators exert their

insecticidal action by disrupting distinct biochemical pathways that are not present in most other organisms, their use as insect control agents must nonetheless be preceeded by extensive studies of their interactions with various components of the environment. A determination of the metabolic and residual fate of these compounds in animals is clearly essential in evaluating their potential toxicological hazards. This paper briefly reviews published reports on the metabolic fate of insect growth regulators in animals, with particular emphasis on preliminary studies conducted in the author's laboratory on the ruminant and insect metabolism of the chitin synthesis inhibitor, diflubenzuron. The reader is also referred to a recent review of the metabolic fate of insect juvenile hormone analogs in insects (Hammock and Quistad, 1976).

METABOLISM OF INSECT JUVENILE HORMONES AND ANALOGS

*Natural Insect Juvenile Hormones*

The metabolism of the natural juvenile hormone (JH) I has been studied in several insect species.

JH 1    $R=R'= -C_2H_5$    (Roller et al., 1967)
JH 2    $R= -CH_3, R'= -C_2H_5$    (Meyer et al., 1968)
JH 3    $R=R'= -CH_3$    (Judy et al., 1973)

The compound was degraded in most insects studied by hydrolysis of the ester group (Slade and Zibbitt, 1972; Whitmore et al., 1972; Ajami and Riddiford, 1972; Weirich et al., 1973; Hammock et al., 1975b; Sanburg et al., 1975; Fox and Massare, 1976), and hydration of the 10,11 epoxide moiety (Slade and Zibbitt, 1972; Ajami and Riddiford, 1973). In at least some insects, the 6,7 double bond was epoxidized and, upon hydration of one of the epoxides, underwent cyclization to tetrahydrofuran diol derivatives (Hammock et al., 1975b). Insects also metabolized JH I to considerable quantities of polar, uncharacterized metabolites, some of which were probably conjugates (Slade and Zibbitt, 1972; Ajami and Riddiford, 1973; Slade and Wilkinson, 1974). JH III was metabolized by insects through at least some of these mechanisms (White, 1972).

Limited studies with mice have shown that orally administered JH I was metabolized to highly polar, unidentified products that were excreted primarily in the urine (Slade and Zibbitt, 1972).

*Phenyl Epoxygeranyl Ether JH Analogs*

Some phenyl 6,7-epoxygeranyl ethers are effective JH mimics and may be useful as insect control agents (Bowers, 1969; Slama, 1971; Pallos et al., 1971). One of the compounds of this group (E)-6,7-epoxy-1-(p-ethylphenoxy)-3,7-dimethyl-2-octene (R-20458 of Stauffer Chemical Company) has been studied extensively with regard to its environmental stability and metabolism by insects and mammals.

Oral administration of phenyl-$^{14}$C labeled R-20458 to mice and rats (Hoffman et al., 1973; Gill et al., 1974) and a steer (Ivie et al., 1976) was followed by essentially quantitative elimination of radiocarbon in the urine and feces within 4 days. In each case, the compound was extensively metabolized with only trace to moderate amounts of intact R-20458 occurring in the feces of the treated animals and none in urine. The major identified metabolic transformations occurring in living rats and mice or in liver microsome systems from rats, mice, and rabbits included hydration of the 6,7-epoxide, 2,3-epoxidation and subsequent hydration, α and β-oxidation of the p-ethyl moiety, ether cleavage and several cyclization reactions (Hoffman et al., 1973; Gill et al., 1974). In the steer, identified metabolic pathways were epoxide hydration, α-oxidation of the p-ethyl moiety, ether cleavage, and reduction of the 6,7-epoxide to an olefin (Ivie et al., 1976). The epoxide reduction was subsequently shown to occur only in the rumen, probably the result of microbial action (Ivie, 1976). R-20458 was also degraded by cockroaches and mealworms through oxidation and hydration reactions that resulted in generally the same metabolic routes as observed in mammals (Hammock et al., 1974b).

## Phenyl 7-Alkoxygeranyl Ether JH Analogs

Several phenyl 7-alkoxygeranyl ethers are potent juvenile hormone mimics (Sarmiento et al., 1973, Hammock et al., 1974a), and the metabolism of two of these compounds, (E)-1-(p-ethylphenoxy)-3,7-dimethyl-7-methoxy or ethoxy-2-octene, has been studied in living insects and in enzyme preparations from insects and mouse liver (Hammock et al., 1975a).

The compounds were metabolized in each of the systems studied primarily by O-dealkylation and α-oxidation of the p-ethyl moiety.

## Alkyl 3,7,11-Trimethyl-2,4-Dodecadienoates

Certain of these compounds are highly active insect JH mimics (Henrick et al., 1973). Methoprene [isopropyl (E,E)-11-methoxy-3,7,11-trimethyl-2,4-dodecadienoate], is currently registered in the United States as a mosquito larvicide and

as a feed through in cattle for control of manure breeding biting flies. The environmental stability and the metabolism of radiolabeled methoprene in several animal and plant species have been reported. The compound was degraded by insects (Solomon and Metcalf, 1974; Quistad et al., 1975a; Yu and Terriere, 1975), cattle and guinea pigs (Chamberlain et al., 1975), chickens (Davidson, 1976; Quistad et al., 1976a) and fish (Quistad et al., 1976b). Ester hydrolysis, O-dealkylation, biological isomerization of the double bond at carbon 2, and carbon-carbon cleavage at the 4,5 double bond were major primary metabolic steps in most of the species studied.

Chickens also produced metabolites resulting from saturation of the dienoate system (Quistad et al., 1976a). A major metabolic pathway for radiocarbon labeled methoprene in mammals was degradation to acetate-$^{14}$C, which was subsequently incorporated into normal body constituents, including glycerides, carbohydrates, proteins, and steroids (Quistad et al., 1974; 1975b,c; 1976a,b).

METABOLISM OF DIFLUBENZURON IN RUMINANTS AND INSECTS

The $N$-(substituted phenyl)-$N'$-(2,6-disubstituted benzoyl) ureas were reported as a new class of insecticides with a unique mode of action by Van Daalen et al. (1972). These compounds interfere with cuticle deposition in insects (Mulder and Gijswijt, 1973; Ishaaya and Casida, 1974; Post et al., 1974), and are extremely toxic to many insect species, yet show very low acute and chronic toxicity to most other animals.

Diflubenzuron [Thompson-Hayward 6040, Trademark Dimilin, $N$-(4-chlorophenyl)-$N'$-(2,6-difluorobenzoyl) urea] is the first compound of this group to undergo commercial development.

Diflubenzuron shows excellent insecticidal activity against several important insect pests and is currently registered in the United States for use against larvae of the gypsy moth. The compound will likely be extensively used in insect control, thus, a thorough evaluation of its environmental fate is needed. To date, only one report dealing with the animal metabolism of diflubenzuron has been published. Metcalf et al. (1975) found that diflubenzuron was quite resistant to degradation by sheep liver microsomes, but did obtain chromatographic evidence for its metabolism to trace quantities of 2,6-difluorobenzoic acid, 2,6-difluorobenzamide, 4-chlorophenyl urea, 4-chloroacetanilide, and 4-chloroaniline and its $N,N$-dimethyl derivative. Some of these products were also seen in snail, mosquito and mosquitofish populations in model ecosystem studies (Metcalf et al., 1975). We have undertaken studies in our laboratory on the metabolic and residual fate of diflubenzuron in ruminants and insects. The remainder of this report considers initial results from these studies.

## Fate of Diflubenzuron-$^{14}$C in Cattle and Sheep

*Treatment and sampling.* A lactating Jersey cow (catheterized) and mixed breed castrate male sheep were held in metabolism stanchions to permit separate collection of urine and feces. In studies designed to measure the elimination of diflubenzuron residues in bile of sheep, the common bile duct was surgically cannulated at least 7 days before diflubenzuron treatment. The radiocarbon labeled diflubenzuron (formulated as a 25% wettable powder (WP) 2-5 µ particle size, with uniform labeling in the two rings and provided by Thompson-Hayward Chemical Co., Kansas City, KS) was diluted with unlabeled diflubenzuron 25 WP, slurried in water, and was administered as single oral doses to the animals *via* stomach tube. The dosage was in each case equivalent to 10 mg/kg of body weight except that in some cases, sheep were given oral diflubenzuron doses of 500 mg/kg to facilitate metabolite isolation in sufficient quantities for spectral analysis. After treatment, urine and feces samples were collected at 24-hour intervals, and the cow was milked every 12 hours. Four to seven days after treatment, the animals were sacrificed and selected tissue samples were collected for analysis of radiocarbon residues.

*Extraction and analysis.* Urine and bile samples were adjusted to pH 2.0, then were exhaustively extracted with ethyl acetate. The extracts were concentrated and then analyzed by thin-layer chromatography (TLC). Whole milk was acidified to pH 2.0 and extracted with ethyl acetate as was the urine. The ethyl acetate extracts were stripped of solvent and the oily residue was partitioned between acetonitrile and hexane. The hexane phase was back extracted once with acetonitrile and the combined acetonitrile fractions were concentrated and analyzed by TLC. Feces samples were homogenized in water, adjusted to pH 2.0, and exhaustively extracted with ethyl acetate. The extracts were then analyzed by TLC.

The radiocarbon content of tissues and feces was determined by oxygen combustion by reported procedures (Ivie et al. 1976). Radioactive residues in liquid phases were quantitated by liquid scintillation counting.

*Metabolite resolution and identification.* Diflubenzuron metabolites in extracts of milk, urine, bile, and feces were resolved by 2-dimensional TLC. The extracts were applied as a spot or short band in one corner of 20 x 20 cm precoated silica gel chromatoplates (0.25 mm gel thickness, with fluorescent indicator, Brinkman Silplate F-22). The plates were developed in benzene-ether (5-1) in the first direction, then in hexane-ethyl acetate-methanol (2-2-1) in the second direction. The developed plates were exposed to Kodak No Screen Medical X-ray film for 4-7 days, then the radioactive areas of the plates were scraped and quantitated by liquid scintillation counting. Metabolite characterizations were made by comparing TLC behavior of radioactive components in the extracts with compounds of known structure (obtained commercially or from Thompson-Hayward) which were suspected to be potential metabolites. In all cases, the metabolite standards were spotted on TLC as a mixture with the $^{14}$C-metabolites, and after 2-dimensional development, the plates were analyzed for coincidence of metabolites and standards. The standards were visualized by viewing the plates under short wavelength ultraviolet light. Coincidence of standard and metabolite constituted tentative metabolite characterization. Isolation of metabolites to obtain sufficient quantities for spectral studies was accomplished by preparative TLC, using 2.0 mm silica gel precoated plates (Brinkman). The plates were developed in either of the two solvent systems used in the 2-dimensional studies, in hexane-ether (2-1), or ether, depending on the metabolite.

Electron impact mass spectral analyses (70 eV) were performed with a Varian-MAT CH-7 magnetic scan spectrometer. Samples were analyzed either by direct insertion probe analysis or through a gas chromatograph coupled with the instrument. The 2 mm x 1.8 m glass column was packed with 3% SE 30 on Varaport 30. Helium flow through the column was maintained at 50 ml/min.

*Radiocarbon excretion and tissue residues.* Radioactive residues were rapidly excreted after oral treatment of the cow and sheep with diflubenzuron-$^{14}$C (Fig. 1). In the cow, about 85% of the administered dose was recovered in the feces during the first 4 days after treatment, about 16% in urine, and only trace residues were secreted into the milk. The intact sheep excreted 41% of the dose in the urine and 43% in the feces during the 4 day posttreatment period. The bile cannulated sheep eliminated 24% of the dose in the urine, 32% in the feces, and 36% in the bile. Analysis of tissues for radiocarbon residues after 4 days (sheep) or 7 days (cow) indicated that only the liver contained appreciable levels of radioactivity, ranging from 2-4 ppm diflubenzuron equivalents.

*Nature of metabolites: Urine.* Two-dimensional TLC of urine extracts resulted in the resolution of 8 radioactive metabolites in the urine of the cow and 9 in sheep (Fig. 2). The relative distribution of metabolites in the urine extracts of bile cannulated and intact sheep was essentially identical.

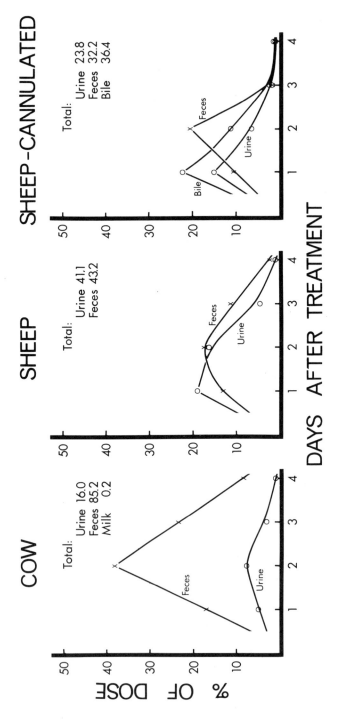

Fig. 1. Elimination of radiocarbon from a cow and sheep treated with a single oral dose, 5.0 mg/kg, of diflubenzuron-$^{14}C$.

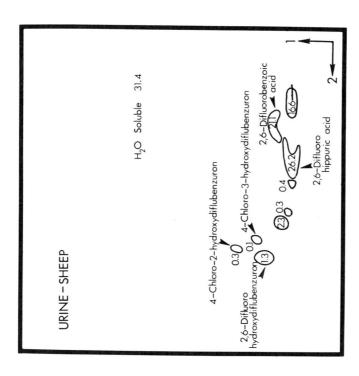
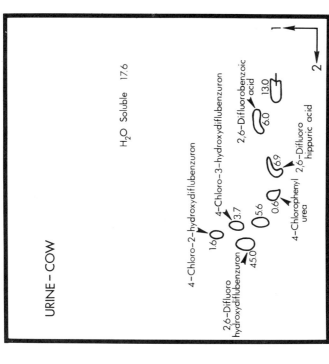

Fig. 2. TLC behavior of diflubenzuron metabolites from cow and sheep urine.

Four of the metabolites occurring in urine were characterized by TLC co-chromatography studies with authentic standards as 2,6-difluorobenzoic acid, 4-chlorophenyl urea, and two-metabolites hydroxylated on the 4-chlorophenyl ring, $N$-(4-chloro-2-hydroxyphenyl-$N$'-(2,6-difluorobenzoyl) urea and its 3-hydroxy isomer. The two hydroxylated metabolites were designated 4-chloro-2-hydroxydiflubenzuron and 4-chloro-3-hydroxydiflubenzuron (Fig. 2). The identifications of 2,6-difluorobenzoic acid and the two isomeric 4-chloro-hydroxydiflubenzuron metabolites were confirmed by mass spectral comparisons with the authentic compounds from synthesis. Two additional metabolites appearing in the urine of both the cow and sheep were identified by mass spectral analysis as 2,6-difluorohippuric acid and $N$-(4-chlorophenyl)-$N$'-(2,6-difluoro-3 or 4-hydroxybenzoyl) urea. The latter metabolite was designated 2,6-difluorohydroxydiflubenzuron (Fig. 2). Fragmentation patterns of the metabolites or their methyl esters as identified by mass spectral analysis are shown in Table 1. No unmetabolized diflubenzuron was seen in urine or either cow or sheep.

Although the same major organic extractable metabolites were seen in the urine of both cow and sheep, there were considerable differences in their relative distribution. The major metabolite in the urine of the cow resulted from hydroxylation of the 2,6-difluorobenzoyl ring and comprised almost half of the radiocarbon in the first day's urine sample (Fig. 2). In contrast, this metabolite was a minor product in sheep urine, in which the major metabolites resulted from cleavage of the amide group at the benzoyl carbon yielding 2,6-difluorobenzoic acid that subsequently was conjugated with glycine to the hippuric acid.

*Bile.* More than one-third of an oral diflubenzuron dose appeared in the bile of a cannulated sheep (Fig. 1), and 2-dimensional TLC resolved the radiocarbon in the ethyl acetate extracts of bile into at least 8 components. Three of these were characterized as the same three hydroxydiflubenzuron isomers seen in urine, but they were minor bile metabolites, collectively comprising <5% of the radioactivity in the bile. Most of the bile radiocarbon was as highly polar compounds; about 50% remained at or very near the origin on TLC and 30% was not extractable from the water phase.

*Feces.* TLC of feces extracts of samples collected 2 days after treatment resulted in resolution into 8 radioactive components from the cow, 6 from the intact sheep, and 1 from the bile cannulated sheep. Unmetabolized diflubenzuron was the major radioactive compound in feces extracts of both cow and intact sheep, but in the cow, 2,6-difluorohydroxydiflubenzuron was a major metabolite that comprised 20-25% of the extracted radioactivity. This metabolite was a minor component (<1%) in the feces extract of intact sheep. Two

TABLE 1

Mass fragmentation patterns of diflubenzuron metabolites from cattle and sheep excreta

| Metabolite | m/e | Ion fragment |
|---|---|---|
| 2,6-Difluorobenzoic acid[1] | 172 | Parent |
| | 141 | Difluorobenzoyl |
| | 113 | Difluorophenyl |
| 2,6-Difluorohippuric acid[1] | 229 | Parent |
| | 197 | Parent $-OCH_3$, H |
| | 170 | Parent $-CO_2CH_3$ |
| | 141 | Difluorobenzoyl |
| | 113 | Difluorophenyl |
| 4-Chloro-2-hydroxydiflubenzuron and 4-Chloro-3-hydroxydiflubenzuron | 326 | Parent |
| | 169 | 4-Chlorohydroxyphenyl isocyanate |
| | 158 | Protonated difluorobenzamide |
| | 157 | Difluorobenzamide |
| | 141 | Difluorobenzoyl |
| | 113 | Difluorophenyl |
| 2,6-Difluorohydroxydiflubenzuron | 326 | Parent |
| | 157 | Difluorohydroxybenzoyl |
| | 153 | Chlorophenylisocyanate |
| | 129 | Difluorohydroxyphenyl |

[1] Analyzed as methyl esters by GLC/mass spectroscopy on a 2 mm x 1.8 m glass column packed with 3% SE 30 on Varaport 30. Column temperature: 100° (methyl difluorobenzoate), 175° (methyl difluorohippurate).

additional metabolites were identified in the feces of both the cow and intact sheep as 4-chloro-2-hydroxydiflubenzuron and 4-chloro-3-hydroxydiflubenzuron, but neither of these metabolites comprised >1% of the radioactivity in the feces samples of either animal. Radioactivity in feces samples from the bile cannulated sheep was readily extractable from the residue phase (>97%) recovery), and TLC revealed the presence of only one radioactive compound that was identified as unmetabolized diflubenzuron.

Milk. The very low levels of radiocarbon secreted into the milk did not permit spectral analysis of metabolites, but three of the seven radioactive components in the milk extracts were characterized by 2-dimensional TLC studies. The major radioactive compound, comprising almost one-half of the total radiocarbon in the milk, was chromatographically identical to diflubenzuron. Two other milk metabolites were identified as 2,6-difluorohydroxydiflubenzuron and 2,6-difluorobenzamide, each comprising about 12% of the total milk radiocarbon.

Fate of Diflubenzuron-$^{14}$C in the Stable Fly and Housefly

Treatment and sampling. Adult stable flies (Stomoxys calcitrans L.) and houseflies (Musca domestica L.) were treated topically on the ventral side of the abdomen with diflubenzuron-$^{14}$C (2 μg in 1 μl acetone/fly). The flies were held in cylindrical plastic holding cups with plastic screen covering each end, and the cages were lined with filter paper to facilitate collection of the excreta. Citrated bovine blood (stable fly) or powdered milk and water (housefly) were provided throughout the study.

Extraction and analysis. Four days after treatment, the flies and the cage lining containing the excreta were separately homogenized in pH 2.0 water, then exhaustively extracted with ethyl acetate. Radioactive products in all extracts were resolved by 2-dimensional TLC.

Absorption, excretion and metabolism. Four days after topical diflubenzuron-$^{14}$C treatment, 40-50% of the applied radiocarbon remained on the treated stable flies, and 20-40% of the dose applied to houseflies was recovered from the treated insects. The excreta contained 35-40% of the dose applied to stable flies and about 50% of that applied to houseflies.

Houseflies were considerably more efficient than stable flies in metabolizing diflubenzuron-$^{14}$C. More than 97% of the radiocarbon recovered from the treated stable flies and their excreta was unmetabolized diflubenzuron, but houseflies degraded about 10% of the dose. Metabolites corresponding to 2,6-difluorobenzamide and 4-chlorophenyl urea were detected in the excreta or whole body extracts of both insect species, whereas trace quantities of 4-chloroacetanilide and 2,6-difluorobenzoic acid were seen in the excreta of stable flies but not houseflies. Based on chromatographic comparisons with metabolites generated in ruminants, neither insect species metabolized diflubenzuron by aryl hydroxylation.

## DISCUSSION

Diflubenzuron-$^{14}$C is metabolized by cattle, sheep, stable flies and houseflies, but the two insect species, particularly stable flies, degrade the compound quite slowly. Diflubenzuron is apparently totally resistant to metabolism by two additional insects, the salt marsh caterpiller (Metcalf et al., 1975) and the boll weevil (Still and Leopold, 1975).

Diflubenzuron is readily metabolized and excreted by cattle and sheep. Once absorbed, the compound apparently requires metabolism to more polar products before excretion because no unmetabolized diflubenzuron appeared in either urine or bile of the treated animals and only trace amounts were secreted into milk. Unmetabolized diflubenzuron was seen in the feces of orally dosed cattle and sheep, but considerable quantities of diflubenzuron metabolites were also seen in the feces. Studies with bile cannulated sheep clearly indicated that the feces metabolites arose through biliary excretion and not by transformations within the digestive tract.

Aryl hydroxylation of the 2,6-difluorobenzoyl moiety was the major metabolic transformation in the cow but was of minor importance in sheep and was not detected in flies. Sheep metabolized diflubenzuron primarily by cleavage of the amide bond at the benzoyl carbon to give 2,6-difluorobenzoic acid that was excreted in the urine either free or conjugated with glycine. The fate of the 4-chloroaniline part of the cleaved diflubenzuron molecule in both ruminants and insects was for the most part not determined because 4-chlorophenyl urea was the single such metabolite identified, and it was detected in only very minor quantity.

## ACKNOWLEDGMENT

I thank Donald Witzel and Wanda Lenger of this laboratory for invaluable assistance in conducting the diflubenzuron ruminant metabolism studies reported here. The diflubenzuron insect metabolism studies were conducted in cooperation with James Wright of this laboratory. Mention of a pesticide in this paper does not constitute a recommendation for use by the U.S. Department of Agriculture nor does it imply registration under FIFRA as amended.

REFERENCES

Ajami, A. M., Riddiford, L. M., *J. Insect Physiol.* 19, 635 (1973).
Bowers, W. S., *Science* 164, 323 (1969).
Chamberlain, W. F., Hunt, L. M., Hopkins, D. E., Gingrich, A. R., Miller, J. A., Gilbert, B. N., *J. Agric. Food Chem.* 23, 736 (1975).
Davison, K. L., *J. Agric. Food Chem.* 24, 641 (1976).
Fox, P. M., Massare, J. S., *Comp. Biochem. Physiol.* 53B, 195 (1976).
Gill, S. S., Hammock, B. D., Casida, J. E., *J. Agric. Food Chem.* 22, 386 (1974).
Hammock, B. D., Gill, S. S., Casida, J. E., *J. Agric. Food Chem.* 22, 379 (1974a).
Hammock, B. D., Gill, S. S., Casida, J. E., *Pestic. Biochem. Physiol.* 4, 393 (1974b).
Hammock, B. D., Gill, S. S., Hammock, L., Casida, J. E., *Pestic. Biochem. Physiol.* 5, 12 (1975a).
Hammock, B. D., Nowock, J., Goodman, W., Stamoudis, V., Gilbert, L. I., *Molec. Cell. Endocrinol.* 3, 167 (1975b).
Hammock, B. D., Quistad, G. B., in, The Juvenile Hormones (L. I. Gilbert, ed.), Plenum, New York, 374 (1976).
Henrick, C. A., Staal, G. B., Siddall, J. B., *J. Agric. Food Chem.* 21, 354 (1973).
Hoffman, L. J., Ross, J. H., Mann, J. J., *J. Agric. Food Chem.* 21, 156 (1973).
Ishaaya, I., Casida, J. E., *Pestic. Biochem. Physiol.* 4, 484 (1974).
Ivie, G. W., *Science* 191, 595 (1976).
Ivie, G. W., Wright, J. E., Smalley, H. E., *J. Agric. Food Chem.* 24, 222 (1976).
Judy, K. J., Schooley, D. A., Dunham, L. L., Hall, M. S., Bergot, B. J., Siddall, J. B., *Proc. Nat. Acad. Sci. USA* 70, 1509 (1973).
Metcalf, R. L., Lu, P. Y., Bowlus, S., *J. Agric. Food Chem.* 23, 359 (1975).
Meyer, A. S., Schneiderman, H. A., Hanzman, E., Ko, J. H., *Proc. Nat. Acad. Sci. USA* 60, 853 (1968).
Mulder, R., Gijswijt, M. J., *Pestic. Sci.* 4, 737 (1973).
Pallos, F. M., Menn, J. J., Letchworth, P. E., Miaullis, J. B., *Nature* 232, 486 (1971).
Post, L. C., de Jong, B. J., Vincent, W. R., *Pestic. Biochem. Physiol.* 4, 473 (1974).
Quistad, G. B., Staiger, L. E., Schooley, D. A., *Life Sci.* 15, 1797 (1974).
Quistad, G. B., Staiger, L. E., Schooley, D. A., *Pestic. Biochem. Physiol.* 5, 233 (1975a).

Quistad, G. B., Staiger, L. E., Bergot, B. J., Schooley, D. A., *J. Agric. Food Chem.* 23, 743 (1975b).
Quistad, G. B., Staiger, L. E., Schooley, D. A., *J. Agric. Food Chem.* 23, 750 (1975c).
Quistad, G. B., Staiger, L. E., Schooley, D. A., *J. Agric. Food Chem.* 24, 644 (1976a).
Quistad, G. B., Schooley, D. A., Staiger, L. E., Bergot, B. J., Sleight, B. H., Macek, K. J., *Pestic. Biochem. Physiol.*, in press (1976b).
Roller, H., Dahm, K. H., Sweeley, C. C., Trost, B. M., *Angew. Chem.* 79, 190 (1967).
Sanburg, L. L., Kramer, K. J., Kezdy, F. J., Law, J. H., *J. Insect Physiol.* 21, 873 (1975).
Sarmiento, R., McGovern, T. P., Beroza, M., Mills, G. D., Redfern, R. E., *Science* 179, 1342 (1973).
Slade, M., Wilkinson, C. F., *Comp. Biochem. Physiol.* 49B, 99 (1974).
Slade, M., Zibbitt, C. H., *Insect Juv. Horm. Chem. Action, Proc. Symp. 1971*, 155-176 (1972).
Slama, K., *Ann. Rev. Biochem.* 40, 1079 (1971).
Solomon, K. R., Metcalf, R. L., *Pestic. Biochem. Physiol.* 3, 127 (1974).
Still, G. G., Leopold, R. A., Paper presented at the 170th National Meeting of the American Chemical Society, Chicago, IL, Aug. 25-29, 1975.
van Daalen, J. J., Meltzer, J., Mulder, R., Wellinga, K., *Naturwissenschaften* 59, 312 (1972).
Weirich, G., Wren, J., Siddall, J. B., *Insect Biochem.* 3, 397 (1973).
Wellinga, K., Mulder, R., van Daalen, J. J., *J. Agric. Food Chem.* 21, 348 (1973a).
Wellinga, K., Mulder, R., van Daalen, J. J., *J. Agric. Food Chem.* 21, 993 (1973b).
White, A. F., *Life Sci.* 11, (Pt II) 201 (1972).
Whitmore, D., Whitmore, E., Gilbert, L. I., *Proc. Nat. Acad. Sci. USA* 69, 1592 (1972).
Yu, S. J., Terriere, L. C., *Pestic. Biochem. Physiol.* 5, 418 (1975).

COMPARATIVE METABOLISM OF SELECTED FUNGICIDES

R. C. Couch and H. W. Dorough

Department of Entomology
University of Kentucky

*ABSTRACT. Fungicides constitute one of the major groups of pesticides in use today and while their total annual production is only about one-third that of either insecticides or herbicides, their chemical, biochemical and toxicological properties are equally as varied and intriguing. From a metabolism standpoint, the fungicides generally undergo reactions in animals that are typical of other xenobiotics. However, many of the organic fungicides are vastly different chemically from other pesticides and drugs, and it is these differences that provide substance for a separate evaluation of the comparative metabolism of fungicides in large animals. Of necessity, rat metabolism data are included for comparative purposes, but unless similar data were published pertaining to nonrodent species for a particular compound, that compound was not included in the present review.*

INTRODUCTION

Of the three major classes of pesticides (insecticides, herbicides, and fungicides), insecticides have been most intensively evaluated with regards to their safety and the subject has been reviewed many times. Herbicides have received somewhat less attention in this regard, but much more than the fungicides. This review will be concerned with the metabolism of selected fungicides in different species and with their general use patterns and toxicological properties.

Metabolic fate studies of any foreign compound are an essential part of the safety evaluation of that compound. They provide evidence of the absorption, distribution, excretion, and retention of the foreign compound and its metabolites in the body. In order for a xenobiotic or its metabolites to become significant to the health of an animal, it must first be available.  If exposure is from an external source, such as in the food or as an environmental contaminant, fate studies are necessary for the evaluation of the degree and extent of exposure.  To further assess the internal exposure it is necessary to determine the metabolic transformations of the foreign compound.  The principal aim is to identify the pathways by which a foreign compound is transformed and to ascertain quantitatively the importance of each pathway.

Problems often arise when attempts are made to extrapolate findings in one species to those in another species.  Interspecies variations originate as a result of difference in one or more of the following:  rate of absorption, mode of excretion, binding to plasma protein and other tissue macromolecules, the rate and route of metabolism, and the nature of the receptor site.  These differences may be due to factors such as different dietary requirements, biliary excretion, gastric pH, intestinal flora, and anatomical differences such as the presence or absence of a rumen (Smith, 1974; Williams, 1971; Anonymous, 1969).

In this presentation the fate of individual fungicides in different species is reviewed, with emphasis on species differences where the data will allow.  Due to the paucity of data dealing with both comparative aspects of fungicide metabolism and toxicology, explanations of species variation in toxicological response based on metabolic differences are nearly nonexistent.

GENERAL CONSIDERATIONS

In terms of total production in the United States of the three major classes of pesticides, fungicides represent only about 15% of the total (The Pesticide Review, 1975).  Sulfur accounts for nearly one-half of this figure.  It is noteworthy that fungicides are the only major class of pesticides that still rely so heavily on inorganics for pest control.  The wood preservative, pentachlorophenol, represents about 32% of the total production of organic fungicides, followed by the dithiocarbamic acid salts at 22% and copper naphthenate at 1%.  The remaining 45% of the organic fungicides consist of the phthalimides (captan), benzimidazoles (benomyl), carboxanilides (carboxin), mercury fungicides (phenyl mercuric acetate), antibiotics (griseofulvin), and others.

The organic fungicides have a diverse chemical makeup.  They contain chemical classes that are traditionally associated with insecticides (organophosphates, carbamates, and organochlorines) and herbicides (nitrophenols and triazines) as well as unique chemical classes such as carboxanilides (carboxin), benzimidazoles (benomyl), phthalimides (captan), piperazines (Triflorine), isoxalones (drazoxolon), naphthoquinones (diclone), and antibiotics (griseofulvin, cyclohexamide).

The mechanisms by which fungicides exert their pesticidal activity are quite varied. Unlike insecticides which are more specific in their action, fungicides and herbicides are more nonspecific in their pesticide behavior. Instead of a single site of action there appears to be several (Lukens, 1971; Moreland, 1967). For example, fungicides exert their activity by interaction with one or more of the following: the cell membrane, mitochondria, the nucleus, ribosomes, respiration, the Krebs cycle, DNA synthesis, or protein synthesis (Lukens, 1971). In addition, the majority of the organic fungicides react indiscriminately with many components and processes of the fungal cell that are common to nearly all living organisms. Their selectivity toward fungi is based primarily upon their rapid uptake and accumulation by the fungal cell (Lukens, 1971).

A review of published acute oral $LD_{50}$ values for organic fungicides (Thomson, 1975), herbicides (Thomson, 1973a), and insecticides (Thomson, 1973b) reveal that organic fungicides and herbicides are not as acutely toxic to vertebrates as are most insecticides. From the data in Table 1 it can be seen that as a class insecticides are by far the most acutely toxic with about 71% having an acute oral $LD_{50}$ value below 500 mg/kg. Fungicides are next in order of toxicity with about 21% having

TABLE 1

*Comparative acute oral toxicity of the three major classes of pesticides in rodents.*

| Toxicity category | Pesticides/% in toxicity category | | |
|---|---|---|---|
| | Insecticides | Fungicides | Herbicides |
| Extremely toxic (1-50 mg/kg) | 27 | 3 | 3 |
| Very toxic (50-500 mg/kg) | 44 | 18 | 10 |
| Moderately toxic (500-3000 mg/kg) | 18 | 34 | 46 |
| Slightly toxic (>3000 mg/kg) | 11 | 45 | 41 |

$LD_{50}$ values below 500 mg/kg. Only 13% of the herbicides are considered in the very toxic and extremely toxic categories.

COMPARATIVE FUNGICIDE METABOLISM

Based on a review of the scientific literature, it appears that less than one-half of the organic fungicides listed by Thomson (1975) have published reports regarding their detailed metabolic fate in animals. While many studies used two or more species, a large proportion dealt only with the residual nature of the fungicides. A review of some of those fungicides (Table 2) that were treated in sufficient detail for a comparative analysis is presented below.

*Benomyl*

Benomyl (Fig. 1, I) is a substituted benzimidazole carbamate used as a systemic foliar fungicide. It effectively controls apple scab, powdery mildew, rots, dutch elm disease, smuts, and many other plant pathogens (Thomson, 1975). It acts as a mite ovicide and also controls certain nematodes (Delp and Klopping, 1968; Millar, 1969). It is used on ornamentals, fruits, mushrooms, field crops, and turf (Farm Chemicals Handbook, 1976). Its acute oral $LD_{50}$ value in male rats is greater than 9,500 mg/kg, with no evidence of chronic toxicity in 90-day feeding tests at a dietary level of 2,500 ppm (Woodcock, 1972). Benomyl in the diet of chickens (25 ppm) for 28 days did not adversely affect feed consumption, body

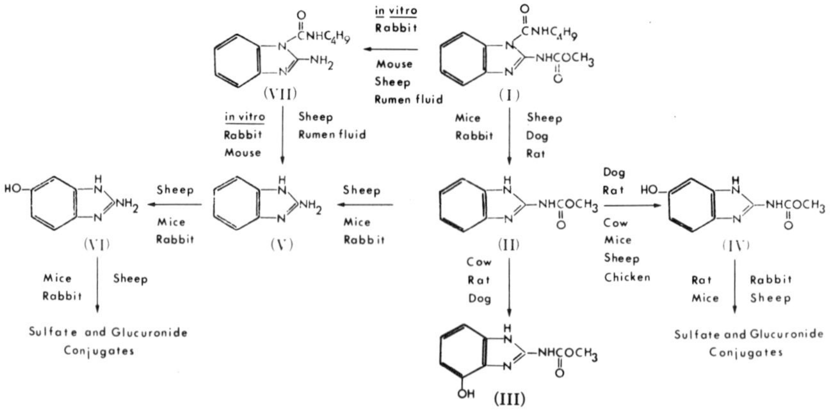

Fig. 1. Metabolism of benomyl (I) in animals.

TABLE 2

*Fungicides considered in the current review.*

| Common Name(s) | Other Name(s) | Chemical Name |
| --- | --- | --- |
| Benomyl | Benlate, Tersan-1991 | Methyl 1-(butylcarbamoyl)-2-benzimidazolecarbamate |
| Furberidazol | Bay 33172 | 2-(2'-furyl)-benzimidazole |
| Thiabendazole | Mertect, Tecto, TBZ | 2-(4-thiazolyl)-benzimidazole |
| Thiophanate-methyl | Cercobin-M, mildothane, Topsin-M | Dimethyl [(1,2-phenylene)bis(iminocarbonothioyl)]-bis[carbamate] |
| Dimethirimol | PP675, Milcurb | 5-n-butyl-2-dimethylamino-4-hydroxy-6-methyl-pyrimidine |
| Drazoxolon | Ganocide, Mil-Col | 4-(2-chlorophenylhydrazono)-3-methyl-5-isoxazolone |
| DSI | -- | N-(3',5'-dichlorophenyl)succinimide |
| Carboxin | Vitavax, D735, DCMO | 5,6-dihydro-2-methyl-1,4-oxathiin-3-carboxanilide |
| PCNB, quintozene | Avicol, Terraclor | pentachloronitrobenzene |
| Hexachlorophene | Hexide, Nabac | 2,2'-methylene bis(3,4,6-trichlorophenol) |
| Griseofulvin | Grisetin | 7-chloro-4,6-dimethoxycoumaran-3-one-2-spiro-1'-(2'-methoxy-6'-methylcyclohex-2'-en-4'one) |

weight gain, or egg production (Gardiner et al., 1974).
Torchinskiy (1973) has described teratogenic and embryotoxic effects of benomyl in rats. However, in a more recent three-generation study using rats, no reproductive or embryonal development differences were observed between control animals and animals receiving dietary levels of benomyl (Sherman et al., 1975). It has been reported to produce chromosomal abnormalities in fungi (Hastie, 1970; Kappas et al., 1974), in the roots of *Allium cepa* (Moll et al., 1973), in various mammalian cell lines, and in specific-pathogen-free rats (Styles and Garner, 1974). Contact dermatitis and bilaterial conjunctivitis were reported to be induced in man by benomyl (Savitt, 1972).

When methyl-2-benzimidazolecarbamate (MBC; Fig. 1, II), a major metabolite of benomyl, was administered intragastrically to Swiss line mice (2 x 250 mg/kg/wk) receiving drinking water containing 0.5% sodium nitrite, there was development of lymphosarcomas in 10 of 30 animals between the 82nd and 164th days of the experiment (Borzsonyi et al., 1975). No developmental anomolies were apparent in offspring of the treated pregnant mice; however, the offspring did later develop lymphosarcomas. Treatment of the mice with either MBC or sodium nitrite-containing water alone did not produce tumors. The implication of the above findings was that a carcinogenic N-nitroso compound was formed *in vivo* from MBC (Borzsonyi et al., 1975). There have been no reports that such reactions occur under practical use situations.

The metabolic fate of benomyl in the rat and dog was studied by Gardiner et al. (1974). After a single oral dose of [$^{14}$C] benomyl was given to a rat and a dog, it was observed that the rat eliminated 79% of the [$^{14}$C] in the urine and 9% in the feces within 72 hr. The dog, on the other hand, excreted 16.2% of the [$^{14}$C] in the urine and 83.4% in the feces. The bulk of the elimination in both species occurred within the first 24 hr. Douch (1973) gave an oral dose of unlabeled benomyl to mice, rabbits, and sheep and collected the urine and feces over a period of 96 hr. In mice 55% of the dose was excreted in the urine and 37% in the feces as metabolites. A similar distribution of metabolites between the urine and feces was found in mice dosed intraperitoneally with benomyl, indicating that biliary excretion may play an important role in the excretion of benomyl metabolites. Sheep and rabbits excreted 58 and 67% of the dose in the urine and 28 and 24% in the feces, respectively. Therefore, a major part of an oral dose of benomyl was absorbed from the gastrointestinal tract and readily excreted in the urine of the different species studied with the exception of the dog. Whether the difference in the route of excretion of benomyl, its metabolites, or both, between dogs and other animals is a reflection of the lack of absorption of benomyl from the gastrointestinal tract or the involvement of significant biliary excretion is not clear. Residue data from studies in rats, dogs, cows, and chickens indicated that benomyl or its major metabolites do not accumulate in animal tissues. Only at high dietary levels of benomyl (25-50 ppm) were residues detectable in the milk from dairy cows and eggs from chickens (Gardiner et al. 1974).

Benomyl was rapidly metabolized by rats, dogs, chickens, cows (Gardiner et al., 1974, 1968), mice, sheep, and rabbits (Douch, 1973). The major metabolic pathway in these animals (Fig. 1) proceeds by way of hydrolytic cleavage of the butylcarbamoyl group with subsequent formation of methyl benzimidazol-2-ylcarbamate (MBC; Fig. 1, II), the predominant metabolite excreted by mice (44% of dose), rabbits (33% of dose), and dogs. Detoxication then proceeds by either hydroxylation of the benzimidazole ring, hydrolysis of the carbamate ester bond, or both. MBC was hydroxylated at the 5(6)-position of the benzimidazole nucleus to form methyl 5-hydroxybenzimidazol-2-ylcarbamate (Fig. 1, IV). In rats and cows this was the principal metabolite and was excreted primarily as the sulfate and glucuronide conjugate. Also, 5-hydroxybenzimidazol-2-ylcarbamate was the only metabolite detected in the excreta and tissues of chickens fed a diet containing 25 ppm benomyl (Gardiner et al., 1974).

Hydroxylation at the 4(7)-position of the benzimidazole nucleus to form methyl 4-hydroxybenzimidazol-2-ylcarbamate (Fig. 1, III) occurred in cows, rats, and dogs, but not in mice, rabbits, or sheep. Sheep, mice, and rabbits cleaved the carbamate moiety of MBC to form 2-aminobenzimidazole (Fig. 1, V). This was the major metabolite excreted in sheep urine and feces (35% of the dose), an amount about twice as much as in mice or rabbits. *In vitro* experiments with sheep rumen fluid (Douch, 1973) indicate that the carbamate ester bond of benomyl may be hydrolyzed to form 1-butylcarbamoyl-2-aminobenzimidazole (Fig. 1, VII) with subsequent formation of 2-aminobenzimidazole prior to absorption. Addition of a hydroxyl group at the 5(6)-position yielded 5-hydroxy-2-aminobenzimidazole (Fig. 1, VI), which was excreted primarily as sulfate and glucuronide conjugates in mice, rabbits, and sheep.

*Fuberidazol*

Fuberidazol (Fig. 2, I), a substituted benzimidazole fungicide, was introduced to replace seed dressing agents containing methylmercury compounds (Frank, 1971). In combination with pentachloronitrobenzene it prevents emergence damage and snow mold in cereal, controls stinking smut of wheat, and also largely prevents soil-borne stinking smut infections (Farm

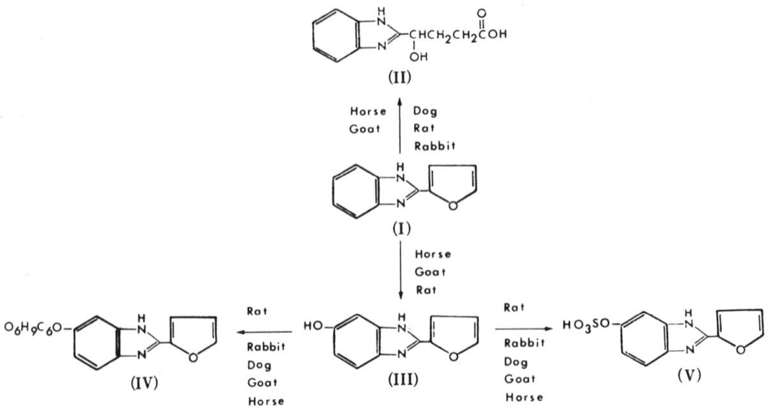

Fig. 2. Metabolism of fuberidazol (I) in mammals.

Chemicals Handbook, 1976). The acute oral, dermal (7-day exposure), and intraperitoneal $LD_{50}$ values of fuberidazol for male rats are 1100, 1000, and 100 mg/kg, respectively. Five doses per week for eight weeks at 120 mg/kg caused an increase in liver weight but had no effect on other organs (Woodcock, 1972). However, it acts as an emetic in certain sensitive dogs (Frank, 1971).

Frank (1971) observed that fuberidazol was rapidly absorbed from the gastrointestinal tract of rabbits, rats, and dogs. Peak blood levels were reached in rats after only 15 min. Appreciable amounts of the fungicide were absorbed from the gastrointestinal tract of horses, goats, and rabbits as reflected in the percentage of the dose excreted in the urine as metabolites 48 hr after an oral dose. Three major metabolites and three to four minor metabolites accounted for 34, 40, and 55% of the dose in the urine of goats, horses, and rabbits. Thus, a principal part of an oral dose of fuberidazol is readily absorbed from the gastrointestinal tract and excreted in the urine of the species studied.

Fuberidazol was metabolized by at least two distinctly different pathways (Fig. 2). One of these involved scission of the furan ring, possibly after hydroxylation of the furan ring, to form 4-(2-benzimidazolyl)-4-hydroxybutyric acid (Fig. 2, II). All of the species studied by Frank (1971) carried out this reaction. However, the ring scission product accounted for 90% of the detectable metabolites found in dog urine after an oral dose of fuberidazol, but in rats, horses, goats, and rabbits, this product accounted for 48, 45, 38 and 35% of the detectable metabolites. The other metabolic pathway involved hydroxylation of the benzimidazol nucleus at the 5(6)-position to form 2-(2-furyl)-5-hydroxybenzimidazole (Fig.

2, III). Trace amounts of the free metabolite were detected in the urine of the horse, goat, and rat, but none was detected in the urine of the rabbit and the dog. The hydroxylated derivative was metabolized further by conjugation with glucuronic acid (Fig. 2, IV) and sulfate (Fig. 2. V). Ratios of the glucuronide conjugate to the sulfate conjugate of the hydroxylated benzimidazole derivative indicate that the horse (4:1) and goat (3:1) preferentially form the glucuronide conjugate while the rat (1:8) and rabbit (1:2) favor conjugation with sulfate. The dog (1:1) shows no preference.

*Thiabendazole*

Originally introduced as an anthelminthic, thiabendazole (Fig. 3, I) has achieved prominence as a broad spectrum systemic fungicide effective against a number of major fungal pathogens (Woodcock, 1972). It controls *Cercospora, Fusarium,* and *Penicillium* diseases of ornamentals, vegetables, and fruit crops (Farm Chemicals Handbook, 1976). The acute oral $LD_{50}$ values for thiabendazole in rats, mice, and rabbits are 3320, 3810, and 3850 mg/kg, respectively. Toxic symptoms were not observed in these species when fed over a two-year period at daily dosage levels of 100 mg/kg (Woodcock, 1972). In dogs, oral doses larger than 200 mg/kg induced vomiting (Robinson et al., 1965). Similarly, thiabendazole may produce nausea in man as well as diarrhea, dermatitis, collapse, tinnitus, and hyperglycemia (Cutting, 1969). Robinson et al. (1965), have suggested that the vomiting in dogs may result from disturbances of the central nervous system.

After oral administration of thiabendazole to sheep (McManus et al., 1966; Tocco et al., 1964), cattle, goats, swine (Tocco et al., 1965), rats, dogs, and man (Tocco et al., 1966), absorption was rapid. The time for attainment of peak

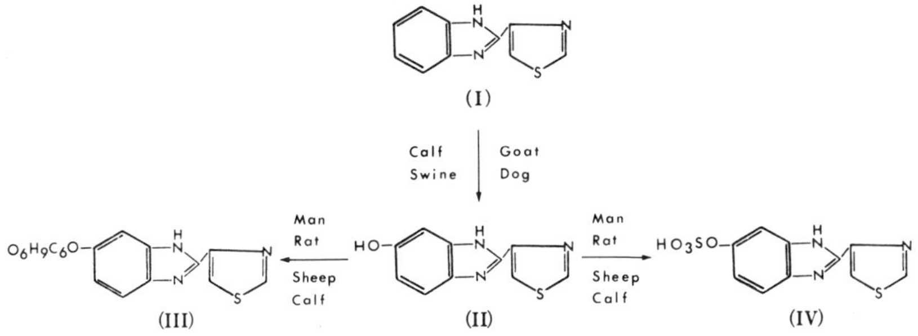

*Fig. 3. Metabolism of thiabendazole (I) in man and other mammals.*

plasma levels after an oral dose of thiabendazole varied between species, with man and dogs showing maximum plasma levels earliest (1-2 hr). Sheep, goats, and cattle took the longest for maximum plasma levels to be attained (4-7 hr), while rats took an intermediary length of time (3-4 hr).

The elimination of thiabendazole and its metabolites from animals receiving an oral dose of the fungicide was rapid. Except for dogs, the urine was the primary route of elimination for thiabendazole and its metabolites. Dogs excreted 35% of an oral dose in the urine and 47% in the feces. Man, on the other hand, excreted approximately 87% of the dose in the urine and only 5% in the feces. More than 40% of the fungicide and its metabolites was excreted in the first 4 hr. Four days after receiving an oral dose of thiabendazole, cattle, swine, goats, and sheep excreted 56, 66, 65, and 75% of the dose in the urine and 30, 10, 23, and 14% in the feces. Therefore, a major part of an oral dose of thiabendazole was absorbed from the gastrointestinal tract of all of the animals studied except for the dog. A greater percentage of the dose may be absorbed than indicated by the percentage of the dose eliminated in the urine and feces. McManus et al. (1966), using the Heidenhain gastric pouch of the goat, showed that thiabendazole and its hydroxylated metabolite (Fig. 3, II) could be secreted back into the gastrointestinal tract after absorption of the parent compound. The possible role that biliary excretion played in the elimination of thiabendazole, its metabolites, or both was not explored in any of the studies reviewed. This route of elimination could be substantial in the dog, where such a large proportion of the dose was eliminated in the feces.

Hydroxylation of the benzimidazole ring at the 5(6)-position to form 5-hydroxythiabendazole (Fig. 3, II) and subsequent conjugation with glucuronic acid (Fig. 3, III) and sulfate (Fig. 3, IV) are major metabolic reactions in the cow, swine, goat, sheep, rat, dog, and man. Data in Table 3 reveal that for the rat, sheep, goat, and swine this pathway accounted for 70-94% of the metabolites appearing in the urine while in the dog and man only 23 and 38% of the urinary metabolites were as these materials. Tocco et al. (1966) suggested that the unidentified metabolites may be the result of ring cleavage or sulfoxidation of the thiabendazole ring. They cite a report by Rosenblum et al. (1964) where it was shown that $^{14}CO_2$ was expired by rats receiving [$^{14}C$]-thiabendazole labeled in the benzene ring. It is also possible that a portion of the unidentified metabolites resulted from scission of the thiazolyl ring. Indeed, as previously indicated, scission of the furan ring, a structure similar to the thiazolyl ring, is a major metabolic reaction in dogs and occurs to a lesser extent in other animals dosed with fuberidazol (Frank, 1971).

TABLE 3

The comparative metabolism of thiabendazole.

| Species | Thiabendazole | 5-Hydroxy thiabendazole | 5-Hydroxy thiabendazole glucuronide | 5-Hydroxy thiabendazole sulfate | Unidentified metabolites | Reference |
|---|---|---|---|---|---|---|
| Rat | 3 | 4 | 28 | 39 | 26 | Tocco et al., 1966 |
| Man | 1 | 1 | 25 | 12 | 61 | Tocco et al., 1966 |
| Sheep | 2 | 10 | 70 | 14 | 4 | Buhs et al., 1964 |
| Cow | 0 | 42 | 21 | 3 | 33 | Tocco et al., 1965 |
| Dog[1] | --- | --- | 23 | --- | 77 | Tocco et al., 1966 |
| Goat[1] | 2 | --- | 70-90 | --- | --- | Tocco et al., 1965 |
| Swine[1] | 2 | --- | 70-90 | --- | --- | Tocco et al., 1965 |

[1] Metabolites were not determined separately.

The nature of the metabolites in the urine (Table 3) indicates that the cow was less efficient in its ability to form sulfate and glucuronide conjugates with 5-hydroxythiabendazole than were the rat, man, or sheep. Conjugation with glucuronic acid was prefered especially by sheep, while rats favored conjugation with sulfate.

*Thiophanate-Methyl*

Thiophanate and thiophanate-methyl (Fig. 4, I) represent a new class of fungicides based on thiourea and containing an aromatic nucleus. Thiophanate is the ethyl analog of thiophanate-methyl. These compounds show a high systemic activity, as well as having curative and preventative properties (Woodcock, 1972). They are used on turf, a number of fruit crops, and small grains. Applied as a foliar spray, soil drench, or seed treatment they control or prevent important diseases such as mildews, rots, and post-harvest decay (Thomson, 1975).

Hashimoto et al. (1972a) have determined the acute oral $LD_{50}$ of thiophahate-methyl in several species. Male mice, rats, guinea pigs, and rabbits had acute oral $LD_{50}$ values of 3.5, 7.5, 3.6, and 2.3 g/kg. Oral administration of 40-500 mg of thiophanate-methyl per day to pregnant mice from day 1 to day 15 of gestation produced no significant cytogenic, mutagenic, or teratogenic effects (Hashimoto et al., 1972b).

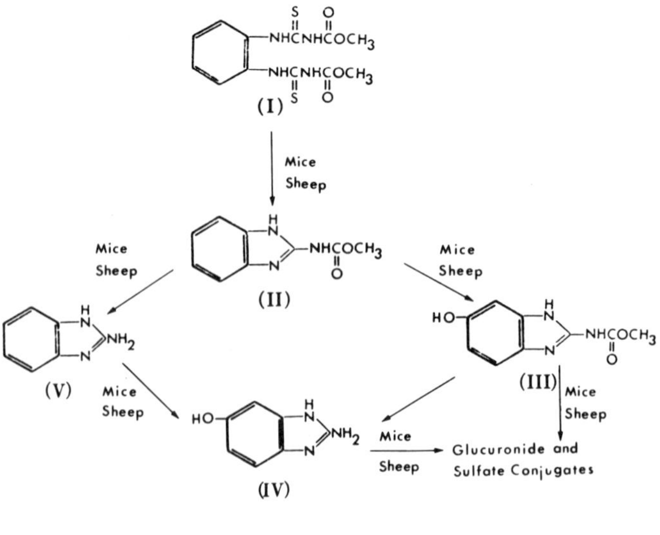

*Fig. 4. Metabolism of thiophanate-methyl (I) in mice and sheep.*

Thiophanate-methyl is readily absorbed from the gastrointestinal tract of mice. Noguchi et al. (1971) orally dosed mice with thiophanate-methyl that was labeled with either a [$^{14}$C]- or [$^{35}$S]-label in one of the thioureido groups. Approximately 80 to 85% of the radioactivity was eliminated in the urine and 17 to 18% in the feces 12 hr after dosing. [$^{14}$C]-Thiophanate-methyl equivalents were rapidly and nearly completely dissipated from tissues of mice given an oral dose of thiophanate-methyl labeled with [$^{14}$C] in a terminal methyl, the benzene ring, or the thioureido group.

Metabolism of thiophanate and thiophanate-methyl in sheep and mice was predominantly by degradative cyclization to benzimidazole derivatives (Fig. 4, II) (Noguchi et al., 1971; Douch, 1974). This was also found to be true in plants (Soeds et al., 1972; Noguchi et al., 1971) and in soil (Fleeker et al., 1974). The conversion may also occur in the absence of enzymes at neutral or alkaline pH (Noguchi et al., 1971). Based on *in vitro* studies with sheep and mice tissues, the cyclization reaction *in vivo* was probably mediated by the mixed-function oxidase enzymes. Metabolism to benzimidazole compounds required NADPH, and was inhibited by SKF-525A and carbon monoxide. In mice and sheep, hydroxylation of the benzimidazole nucleus at the 5(6)-position occurred (Fig. 4, III, IV) as did hydrolysis of the carbamate moiety with subsequent formation of 2-aminobenzimidazole (Fig. 4, V) (Douch, 1974).

Seventy-two hr after receiving an oral dose of unlabeled thiophanate-methyl, mice excreted 59% of the dose in the urine and feces as the cyclized metabolites shown in Fig. 4, while sheep excreted 42% of the dose as these metabolites via the urine and feces. Since the presence of other metabolites in the excreta was not reported, it is unclear whether this reflected a greater activity of a minor pathway(s) in mice or was indicative of a slower excretion rate in sheep. Indeed, sheep and mouse liver preparations, as well as rumen fluid, gave six to seven metabolites in addition to the major ones when incubated with thiophanate or thiophanate-methyl. The minor metabolites resulted from replacement of sulfur with oxygen and/or cleavage of a methoxy- or an ethoxycarbonyl moiety in one of the thioureido groups. Douch (1974) stated that the pattern of metabolites formed *in vitro* were the same, but gave no quantitative data that would permit an assessment of the relative importance of the minor pathway(s) in the two species.

*Dimethirimol*

Dimethirimol (Fig. 5, I) is a systemic fungicide containing a pyrimidine nucleus. It is effective as a soil and spray treatment for the control of powdery mildew on cucumbers,

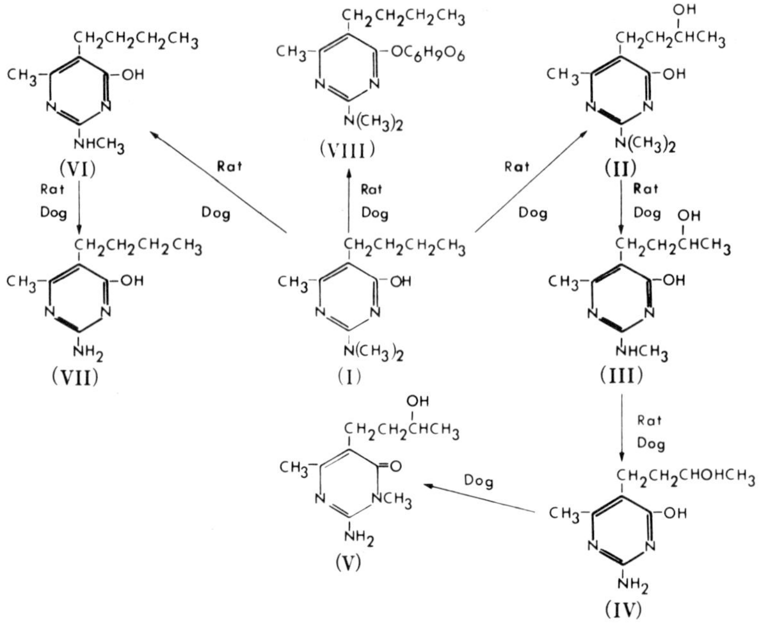

Fig. 5. Metabolism of dimethirimol (I) in rats and dogs.

melons, and certain ornamentals (Farm Chemicals Handbook, 1976). The acute oral $LD_{50}$ of dimethirimol in rats is greater than 4000 mg/kg and a three-day dermal application of 400 mg/kg produced neither skin damage nor toxic symptoms. Dietary levels of 1000 ppm for 90 days produced no adverse effects in rats or dogs (Woodcock, 1972). The 2-year no-effect levels in rats and dogs correspond to an intake of 300 ppm in the diet and 25 mg/kg per day, respectively (Bratt et al., 1972).

Dimethirimol was readily and nearly completely absorbed from the gastrointestinal tract of rats and dogs as evidenced by the fact that 76 and 85% of an oral dose of $[^{14}C]$-dimethirimol labeled in the heterocyclic ring was excreted in the urine within the first 24 hr after an oral dose (Bratt et al., 1972). After three days nearly all of the radiolabel could be accounted for in the urine and feces. No $^{14}CO_2$ was detected in the expired air of rats during this time period, suggesting that the heterocyclic ring remained intact in this species. Since a similar $^{14}CO_2$ trapping experiment was not conducted for dogs receiving an oral dose of $[^{14}C]$-dimethirimol, the possibility that ring scission may have occurred cannot be ruled out. That enterohepatic circulation may play an important role in the fate of dimethirimol in rats was indicated by the observation that, given an intraperitoneal dose of [$^{14}$C]-dimethirimol, rats excreted 89% of the radiolabel in the urine and 11% in the feces over the same time period as the oral dosing experiment (Bratt et al., 1972). This was further substantiated when the bile ducts of orally dosed rats and dogs were cannulated. Rats excreted approximately 28% of the dose in the bile within the first 48 hr, with the greater proportion excreted within the initial 24 hr. It was estimated that dogs excreted 37% of the dose in the bile in 24 hr, with the greater proportion being excreted within 6 hr after dosing.

The major route of metabolism of dimethirimol in both the rat and the dog was *via* sequential dealkylation of the dimethylamino group and penultimate hydroxylation of the n-butyl substituents (Fig. 5, II-VII) (Bratt et al., 1972). Metabolites formed by this route accounted for 75 and 54% of the total metabolites appearing in the rat and the dog urine. Dogs appeared able to carry the sequence of reactions one step further and form 4-(2-oxo-3-*N*-methyl-4-amino-6-methylpyrimidinyl)butan-2-ol (Fig. 5, V). The only conjugated material identified was the glucuronic acid conjugate of dimethirimol (Fig. 5, VIII) and was found predominately in the bile. Rats excreted a larger proportion of the dose in the bile as this metabolite than did the dogs (Calderbank, 1971; Bratt et al., 1972). While about 2% of the dose was identified as the glucuronic acid conjugate in the urine of rats, none was detectable in the urine of dogs (Bratt et al., 1972). Even though the major metabolites excreted by the rat and dog were similar, differences in minor metabolic pathways may exist since the bile of dogs contained 10 radiolabeled compounds and rat bile contained only 6 radiolabeled compounds.

*Drazoxolon*

Drazoxolon (Fig. 6, I) is one of a series of arylazoisoxazolones that have been found to give effective control of powdery mildew and a number of soil and seed-borne plant diseases (Farm Chemicals Handbook, 1976). It is sold outside the U.S., where it is used on rubber trees, coffee, and tea and as a seed treatment of peas, beans, corn, peanuts, and cotton (Thomson, 1975). It is more toxic than most fungicides and shows an apparent species difference in acute oral toxicity. The $LD_{50}$ value in rats is 125 mg/kg, while in dogs the $LD_{50}$ value is 17 mg/kg (Daniel, 1969).

Approximately 75% of a single oral dose of [$^{14}$C]-drazoxolon, labeled in the isoxazol ring, was excreted in urine and 13% in the feces of rats after 96 hr (Daniel, 1969). An additional 7% of the dose was expired as $^{14}CO_2$. Bile duct-cannu-

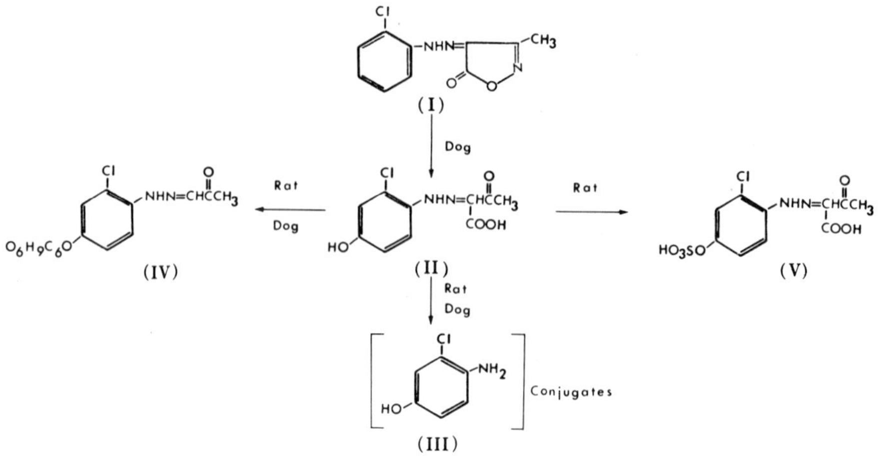

Fig. 6. Metabolism of drazoxolon (I) in rats and dogs.

lated rats that received an oral dose of [$^{14}$C]-drazoxolon excreted about 14% of the dose via the bile. When dogs were given an oral dose of [$^{14}$C]-drazoxolon about 35% of the label was excreted in the urine and a similar amount in the feces after 96 hr. The contribution, if any, that the bile played in the excretion of drazoxolon from the dog was not determined.

In both the rat and the dog, scission of the isoxazolone moiety is a major metabolic reaction. Hydroxylation of the phenyl ring at the 4-position may precede hydrolysis, as no evidence was obtained for the presence of 2-(2-chlorophenyl-hydrazono)acetoacetic acid in the excreta of the rat or dog (Daniel, 1969). In rats, 2-(2-chloro-4-hydroxyphenylhydrazono)acetoacetic acid (Fig. 6, II) was excreted in the urine conjugated with sulfate (Fig. 6, III) and glucuronic acid (Fig. 6, IV). In contrast, dogs eliminated in about equal proportions the free metabolite (Fig. 6, II) and its glucuronide conjugate (Fig. 6, IV). While the sulfate conjugate of 2-(2-chloro-4-hydroxyphenylhydrozono)acetoacetic acid was the primary metabolite found in the urine of rats, none was detectable in the urine of dogs. An undefined conjugated form(s) of 4-amino-3-chlorophenol (Fig. 6, V) was detected in the urine of dogs and rats. Rats appeared to be slightly more efficient in hydrolyzing the N-N bond. Also, rat urine contained an unknown metabolite(s) that was not detectable in dog urine. Identification of metabolites in the feces was not determined and tissues were not analyzed for residues. As much as 75% of an oral dose of drazoxolon given to dogs was not characterized. For rats, the situation was better in that only about 12% of the oral dose remained to be accounted for by the spectrophotometric and chromatographic procedures used.

## N-(3',5'-Dichlorophenyl)succinimide

The fate of the experimental fungicide N-(3',5'-dichlorophenyl)succinimide (DSI; Fig. 7, I) in rats and dogs was studied by Ohkawa et al. (1974). DSI, one of the most active fungicides of the cyclic imide group, is effective against a broad spectrum of plant pathogenic fungi, especially against *Sclerotinia sclerotiorum* and *Botrytis cinerea*. The acute oral $LD_{50}$ values in mice and rats are 1350 and 890 mg/kg (Ohkawa et al., 1974).

Peak levels of radioactivity were found in rat blood 4 hr after oral administration of [phenyl-$^3$H] DSI (Ohkawa et al., 1974). Male rats orally dosed with [$^{14}$C]-DSI excreted approximately 94% of the radioactivity either in urine (69.7%), feces (17.5%), or expired air (5.9%) within the first 48 hr. That DSI was incompletely absorbed from the gastrointestinal tract of male dogs was suggested by the observation that about 28% of an oral dose of DSI was excreted unchanged in the feces. An average of 71.1% of the dose was excreted, with 32.4% in the urine and 39.7% in the feces 48 hr after dosing. Unfortunately, expired [$^{14}$C]-gases were not trapped, nor were tissue residues determined; therefore, about 30% of the dose was not accounted for in the dog. Whole body radioautograms of rats showed high levels of radioactivity in the stomach and intestine 30 min after oral administration of [$^{14}$C]-DSI. After 4 hr, especially high radioactive levels were found in the kidney, lung, liver, and intestines. Twenty-four hr after dosing, this method detected no [$^{14}$C]-residues in the animal, thus indicating rapid elimination of DSI, its metabolites, or both from the animal.

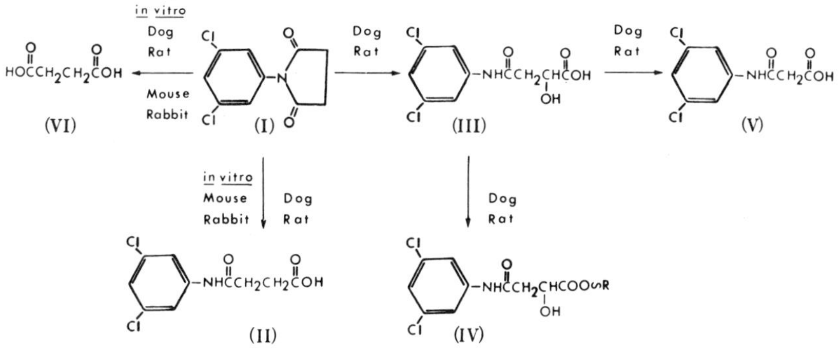

Fig. 7. Metabolism of N-(3',5'-dichlorophenyl) succinimide in rats and dogs.

The metabolism of DSI by rats and dogs was qualitatively similar, although the proportion of metabolites was different. The primary urinary metabolites were identified as $N$-(3',5'-dichlorophenyl)succinamic acid (Fig. 7, II), $N$-(3',5'-dichlorophenyl)malonamic acid (Fig. 7, V), $N$-(3',5'-dichlorophenyl)-2-hydroxysuccinamic acid (Fig. 7, III), and $N$-(3',5'-dichlorophenyl)-2-hydroxysuccinamic acid derivatives (Fig. 7, IV). The $N$-(3',5'-dichlorophenyl)-2-hydroxysuccinamic acid derivatives (Fig. 7, IV) were the main urinary metabolites in dogs representing 66.8% of the total radiocarbon in urine; in rat urine they accounted for only 26.8% of the total radiocarbon. While the DSI metabolites, $N$-(3',5'-dichlorophenyl)succinamic acid (rat, 19.9%; dog, 1.5%), $N$-(3',5'-dichlorophenyl)-2-hydroxysuccinamic acid (rat, 15.0%; dog, 3.3%), and $N$-(3',5'-dichlorophenyl)malonamic acid (rat, 11.9%; dog, 3.4%), were proportionately higher in rat urine than in dog urine, the unknown metabolites (rat, 25.6%; dog, 24.5%) in the urine were proportionately the same. The nature of the $N$-(3',5'-dichlorophenyl)-2-hydroxysuccinamic acid derivatives in both rats and dogs could not be clearly defined. Treatment of these metabolites with β-glucuronidase or aryl sulfatase did not yield any aglycones. Results of derivatization of this fraction with diazomethane suggested that the chemical nature of the acid derivatives isolated from dog urine was different than those obtained from rat urine.

In vitro studies with dog, mouse, rat, and rabbit liver homogenates demonstrated that DSI was metabolized similarly. However, the rabbit most effectively converted DSI to $N$-(3',5'-dichlorophenyl)succinamic acid and succinic acid (Fig. 7, V). Dog liver homogenate had the lowest activity while mouse and rat liver were of about equal activity and were twice as active as dog liver, but about one-half that of rabbit liver (Ohkawa et al., 1974).

*Carboxin*

Carboxin (Fig. 8, I), an oxathiin carboxanilide, is a systemic fungicide used as a seed treatment for wheat, oats, cotton, and barley seeds to control smuts, bunts, *Rhizoctonia spp.*, damp off, and common bunt (Thomson, 1975). The acute oral $LD_{50}$ of carboxin in rats is about 3820 mg/kg and the acute dermal $LD_{50}$ in rabbits is greater than 8000 mg/kg. Rats fed 600 ppm carboxin daily in their diet for 2 years suffered no detectable symptoms (Martin, 1972). Stefan et al. (1973) observed decreased erythrocyte and increased leukocyte counts in rats treated with carboxin (0.015 mg in the diet/day) for 30-45 days. Carboxin decreased serum total protein and glutamic-oxalacetic transaminase but increased $\alpha_1$- and $\alpha_2$-glob-

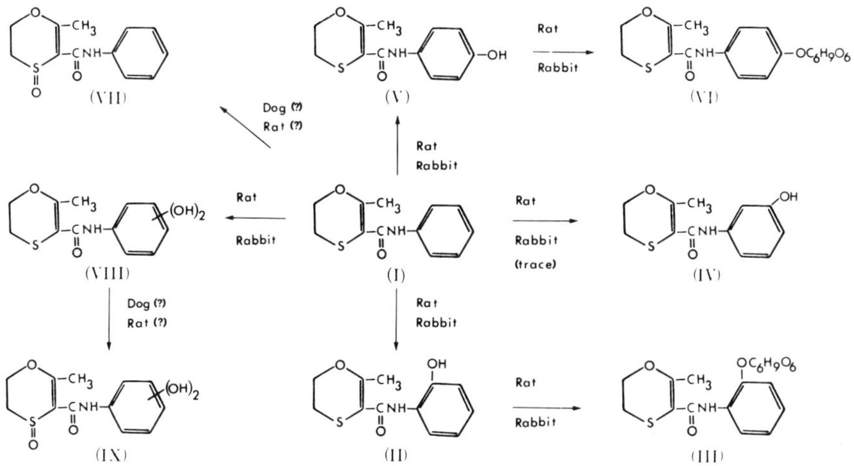

Fig. 8. Metabolism of carboxin (I) in animals.

ulin, glutamic-pyruvic transaminase, and alkaline phosphatase serum levels.

Carboxin was almost completely absorbed from the gastrointestinal tract of rabbits, but much less so from the gastrointestinal tract of rats (Waring, 1973). Ninety percent of an oral dose of [$^{14}$C]-carboxin was eliminated in the urine of rabbits 24 hr after dosing. Rats, on the other hand, excreted about 50% of the oral dose primarily as the unmetabolized fungicide in the feces over the same time period. Bile cannulation experiments in rats indicated that biliary excretion did not contribute to the high levels of parent fungicide in the feces. Only trace amounts of p-hydroxy carboxin (Fig. 8, V) and its glucuronide conjugate (Fig. 8, VI) were found in the bile (Waring, 1973). Gastric and/or salivary secretion of the fungicide was not ruled out as contributing to the large percent of parent compound found in the feces. In this regard, autoradiograms of rats 2 hr after receiving carboxin labeled in either the heterocyclic or the aromatic ring showed radioactivity to be localized in the salivary glands, the intestinal tract, and the liver. Six hr after dosing radioactivity was also present in the kidney. However, 48 hr after dosing only trace amounts of [$^{14}$C] were detectable in any of the tissues of the rat. Quantitative data on the distribution of radioactivity in the tissue of the rabbit were not reported.

Chin et al. (1971) analyzed tissues of dogs fed carboxin for two years at 600 ppm in the diet and observed that there were 5 ppm carboxin residues in the liver, 0.5 ppm in the kidney, and no residues in the fat or muscle tissues. Analysis of the excreta showed an elimination pattern similar to rats.

Fig. 8 outlines the proposed metabolic pathways for carboxin metabolism in animals. Hydroxylation of the aromatic ring moiety (Fig. 8, II, IV, V, VIII) and subsequent conjugation with glucuronic acid (Fig. 8, III, VI) represents the major metabolic pathway in both rabbits and rats (Waring, 1973). Rabbits hydroxylated 88% of the dose while rats hydroxylated 39% of the dose. Rabbits conjugated the phenol with glucuronic acid to a greater degree than the rat. The ratio of glucuronide conjugated metabolites to free phenol in the urine of rabbits was 28:1 and only about 5:1 in the rat urine. Quantities of unmetabolized carboxin excreted in the urine of rats (15% of the dose) as compared to that in the rabbit (2% of the dose) suggested that rats do not metabolize the fungicide as readily as do rabbits.

Chin et al. (1971) reported that sulfoxidation of carboxin was a significant metabolic reaction in dogs (Fig. 8, VII, IX). Oxidation to the sulfoxide was also reported to be the principle reaction of carboxin in soil, plants, and fungi (Chin et al., 1971; Wallnoefer et al., 1972; Leroux and Gredt, 1972; Lyr et al., 1974). It should be noted, however, that Briggs et al. (1974) showed that the major metabolite isolated from barley plants treated with carboxin was the p-hydroxyphenyl derivative and it has been suggested (Waring, 1973) that the TLC method used to identify the sulfoxidation metabolites from dogs, soil, and plants did not readily discriminate between the sulfoxide and the phenol. Either or both may have been present. Therefore, additional work is needed to clarify this apparent species difference in carboxin metabolism.

## Pentachloronitrobenzene

Pentachloronitrobenzene (PCNB; Fig. 9, I) is a chlorinated hydrocarbon used as a soil fungicide and seed disinfectant. It is a broad spectrum fungicide that controls or prevents a number of important diseases such as wheat bunt, snow mold, Rhizoctonia spp., Sclerotina spp., potato scab, black rot, and Southern blight. It is used on a wide range of vegetable crops, ornamentals, and turf (Thomson, 1975). The acute oral toxicity of PCNB in rats is 12,000 mg/kg (Martin, 1972) and the acute percutaneous $LD_{50}$ of PCNB in rabbits is greater than 4000 mg/kg (Borzelleca et al., 1971). No adverse reproductive effects appeared in rats fed dietary concentrations of PCNB as high as 500 ppm. Two-year feeding studies in dogs established a no-effect level of 30 ppm. Cholestatic hepatosis with secondary bile nephrosis was found at the 180 ppm level in the diet (Borzelleca et al., 1971). PCNB was nonmutagenic in drosophilia (Vogel and Chandler, 1974) and bacteria (Seiler,

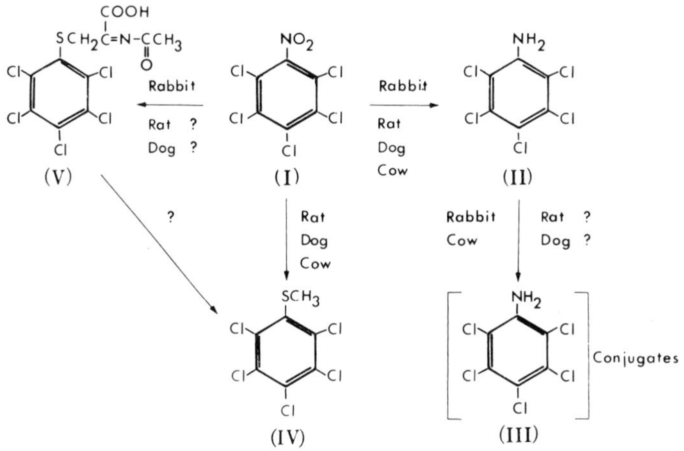

Fig. 9. Metabolism of pentachloronitrobenzene (I) in mammals.

1973). However, it has been reported to produce hepatomas in mice (Innes et al., 1969). Teratogenicity was demonstrated in mice (Courtney, 1973) but not in rats (Jordan and Borzelleca, 1973).

PCNB shows variability in its absorption from the gastrointestinal tract of dogs, rats (Kuchar et al., 1969; Borzelleca et al., 1971), rabbits (Betts et al., 1955), and cows (St. John et al., 1965). Rabbits excreted more than 50% of an oral dose of PCNB unchanged in the feces, suggesting that PCNB was poorly absorbed by this species (Betts et al., 1955). Balance studies were not conducted in the other species, thereby making it impossible to assess the extent of absorption of PCNB from the gastrointestinal tract. However, in cows PCNB was absorbed to the extent that 45% of the dose was excreted in the urine as the metabolite, pentachloroaniline (Fig. 9, II). Twenty-four hr urine and feces samples from dogs fed 1080 ppm PCNB in their diet for two years indicated that PCNB was not completely absorbed. Unmetabolized PCNB (14 ppm) along with its major metabolites pentachloroaniline (16.7 ppm) and methyl pentachlorophenyl sulfide (Fig. 9, IV) (3.64 ppm) were detected in the feces. The corresponding values for the parent compound and major metabolites in urine were as fractional ppm. The bile was free of intact PCNB, but did contain pentachloroaniline (1.9 ppm) and methyl pentachlorophenyl sulfide (4.5 ppm) (Borzelleca et al., 1971). Rats fed a diet containing 500 ppm PCNB excreted no detectable PCNB in the feces, but did excrete trace amounts of pentachloroaniline and methyl pentachlorophenyl sulfide (Borzelleca et al., 1971).

There was little storage of PCNB residues in the tissues

of rats, dogs, or cows upon chronic exposure to this fungicide. Most of the PCNB equivalents in the tissues were as pentachloroaniline and methyl pentachlorophenyl sulfide (Borzelleca et al., 1971). Residue levels in the rabbit were not reported.

Fig. 9 shows a proposed metabolic pathway of PCNB in animals which demonstrates that PCNB is metabolized largely *via* reduction of the nitro group with subsequent formation of the aniline derivative, pentachloroaniline. Acid hydrolysis of cow milk, kidney, and liver residues resulted in an increase in pentachloroaniline, indicating that conjugates were formed (Borzelleca et al., 1971). The nature of the conjugate was not defined. Rabbits also excreted pentachloroaniline in the urine possibly as a glucuronic acid or sulfate conjugate (Fig. 9, III) (Betts, 1955).

The replacement of the nitro group with a methylthio group to form methyl pentachlorophenyl sulfide was an important metabolic reaction in rats, dogs, and cows (Kuchar et al., 1969; Borzelleca et al., 1971). This unique reaction apparently also occurs in soil treated with PCNB (deVos, et al., 1974). Rabbits metabolized PCNB by replacement of the nitro group to form pentachlorophenyl mercapturic acid (Fig. 9, V) (4-14% of the dose), but it was not specifically assayed for in the other species studied and no comparison could be made (Betts, 1955). The studies conducted thus far with PCNB have been with unlabeled material and have failed to account for all the compound given and therefore other metabolic pathways may exist.

*Hexachlorophene*

Hexachlorophene (Fig. 10, I) is a chlorinated bisphenol used as a soil fungicide. It is used on cotton, cucumbers, tomatoes, and peppers to control damping off, bacterial spot and a number of other soil borne diseases. It is a broad-spectrum fungicide with chemotherapeutic action (Farm Chemicals Handbook, 1976). Kimbrough (1976) has recently reviewed the toxicity of this compound and reports acute oral $LD_{50}$ values in adult sheep, rats, cattle and mice of 30, 56, 60, and

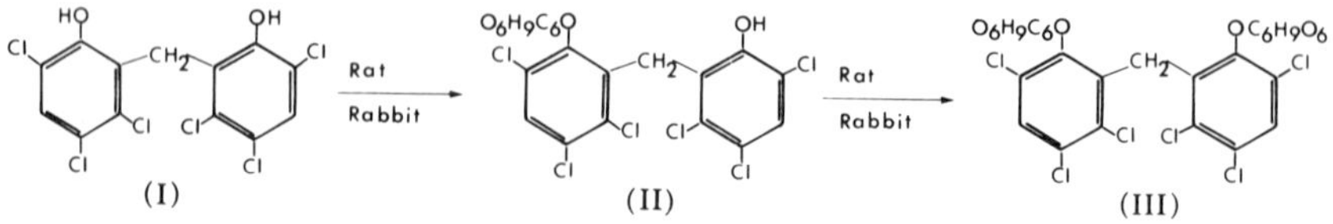

Fig. 10. Metabolism of hexachlorophene (I).

168 mg/kg. The $LD_{50}$ varied markedly in the rat, depending on its stage of development. In the suckling rat the $LD_{50}$ was 9 mg/kg, but in weanling rats, the $LD_{50}$ was 120 mg/kg.

A most striking toxic response was degenerative change of the white matter of the brain and spinal cord, known as status spongiosus. This toxic manifestation may be precipitated in infants, monkeys, sheep, mice, rabbits, cats, and tadpoles after application of hexachlorophene to the skin or incorporation of the fungicide into the diet. Young animals are more susceptible than are the adults. However, it should be pointed out that Kennedy et al. (1976) have shown that such changes produced by hexachlorophene in rats are reversible following discontinuance of treatment. Sheep are unique in that they show an irreversible lesion of the optic nerve not seen in other species studied. Introduction of high doses of hexachlorophene into the vagina of pregnant rats produced hydrocephalus in the rat fetus. Also, the survival rate of rat pups was reduced when hexachlorophene was incorporated into the dams diet (Kimbrough, 1976).

Gandolfi and Buhler (1974) reported that a rapid and almost complete absorption of an oral dose of hexachlorophene occurred in rats. In fact, following either an oral or intraperitoneal dose of $^{14}C$-hexachlorophene, it was said that equivalent levels of radioactivity appeared in the tissues within 2 hr after administration. The specific levels were not given. The primary route of excretion was apparently via the bile and into the feces (52-75% of the dose) following intraperitoneal administration of $[^{14}C]$-hexachlorophene to rabbits and rats.

Hexachlorophene was readily absorbed from normal and burned skin of man (Ulsamer et al., 1973). Blood levels after dermal application to normal skin ranged from 0.38 µg/ml in an individual using a 3% hexachlorophene cleanser for whole-body washing to less than 0.005 µg/ml (limit of detection) in some individuals using a hexachlorophene cleanser for hand-washing only. Wit and VanGenderen (1962) reported that rats and rabbits receiving an oral dose of $[^{14}C]$-hexachlorophene had excreted about 70% of the radiolabel in the feces within a week after receiving the fungicide. About 32% of the dose was excreted unchanged in the feces of rabbits and about 35% was as an unknown metabolite. Only about 2% of the dose was excreted in rat urine, but rabbit urine contained about 31% of the dose primarily as unchanged hexachlorophene. Cows, like rats, excrete an oral dose of hexachlorophene via the feces (64%) and not the urine (Wit and Van Genderen, 1962; St. John and Lisk, 1972).

Three days after an intraperitoneal dose of $[^{14}C]$-hexachlorophene to rats, 30-47% of the dose was eliminated via the bile (Gandolfi and Buhler, 1974). Ligation of the bile-duct increased the toxicity of hexachlorophene to rats and changed the major route of excretion to the urine (55% of the dose). The principle metabolite in bile duct-ligated rats was hexachlorophene monoglucuronide.

Mitin et al. (1969) gave an alcohol solution of $[^{14}C]$-hexachlorophene to pregnant guinea pigs and found that 12% was voided in the feces and 7% in the urine after 24 hr. In the organs, the distribution of radioactivity expressed as % of dose after 24 hr was 2.2% in the liver and in the thyroid, 1.4% in the kidney, 1.6% in the blood, and 0.76% in the spleen. In the fetal liquid the activity was equivalent to 0.52% of the dose and in the fetal liver, 0.99%. They also reported that goats given an oral dose of $[^{14}C]$-hexachlorophene excreted radioactivity in the feces and milk over a 5-day period at the end of which no radioactivity was detectable in the milk. This contrasts with observations in cows where no hexachlorophene equivalents were detectable in the milk (St. John and Lisk, 1972; Wit and Van Genderen, 1962).

Information on the metabolic fate of hexachlorophene in animals is limited. Hexachlorophene monoglucuronide (Fig. 10, II) was identified as the major metabolite of hexachlorophene in the urine of rats and rabbits. This was also the major metabolite found in the bile of rats. The diglucuronide of hexachlorophene (Fig. 10, III) was also a urinary metabolite in rats and rabbits (Gandolfi et al., 1972).

*Griseofulvin*

Griseofulvin (Fig. 11, I) is an antibiotic used as a foliar fungicide on various vegetable crops in Europe but not in the United States. It controls or prevents grey mold (*Botrytis*) of lettuce, powdery mildew and a few other plant pathogenic fungi (Thomson, 1975). In the United States it has found therapeutic uses in the treatment of mycotic diseases of the skin, hair, and nails due to *Microsporum*, *Trichophyton*, and *Epidermophyton*. Griseofulvin is also highly effective in treatment of "athlete's foot" (Goodman and Gilman, 1970). It is one of the most effective oral antifungal agents now available. It is relatively nontoxic, although headache, peripheral neuritis, nausea, and mild rashes have been observed in man. In experimental animals, large doses depress the hematopoietic system and interfere with mitosis and spermatogenesis (Cutting, 1969).

An excellent review dealing with the metabolic fate of griseofulvin in man and animals has been written by Lin and Symchowicz (1975). The following comments will serve to summarize the species differences in the metabolic fate of griseofulvin.

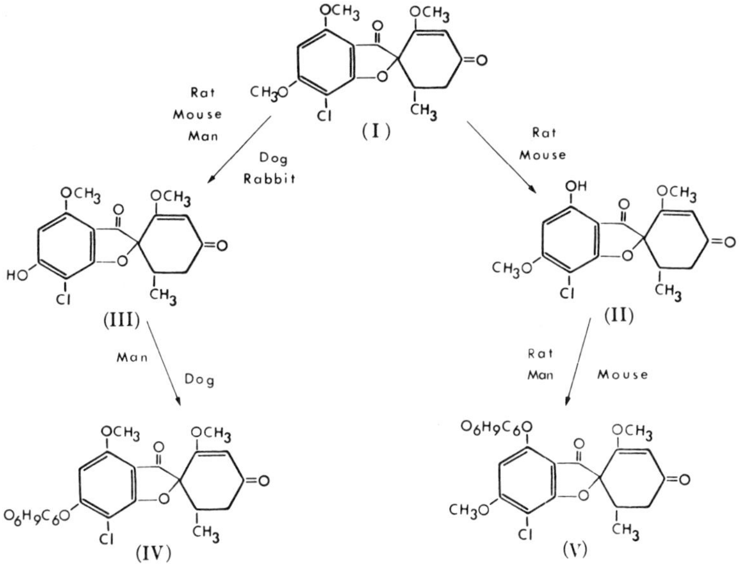

Fig. 11. Metabolism of griseofulvin (I) in man and other mammals.

Griseofulvin is variably and incompletely absorbed after oral administration to man and laboratory animals with a number of factors influencing its absorption such as particle size, fat intake in the diet, formulation, dosage regimen, and species. With regard to the latter, after an oral dose of micronized griseofulvin, the percentages of the dose absorbed in man and dog were 44 and 33%, respectively. When griseofulvin was given orally as an aqueous suspension, 60 and 85% of the doses were absorbed by the rat and mouse (Lin and Symchowicz, 1975). In addition to species differences, there is also a wide subject-to-subject variation in the absorption of griseofulvin. Rowland et al. (1968) reported that the amount of fungicide absorbed by man may vary from 27 to 73% of the dose. Humans dosed with 0.5 g of micronized griseofulvin absorbed approximately 50% of the dose in the first 3-10 hr. In the above studies the dose absorbed was calculated by dividing the percentage of griseofulvin equivalents excreted in the urine after oral administration by that excreted after intravenous administration. The rate of absorption in rats was similar to that observed in man with peak plasma levels being attained within 3 to 7 hr after oral dosing with several different dosage forms (Carrigan and Bates, 1973).

After intravenous administration the plasma half-life of griseofulvin shows marked species differences. The plasma half-life for the first or fast component was 30, 20, 7, and 4 min, respectively, in the mouse, rat, rabbit and dog. Man exhibited the longest plasma half-life which varied from 42 to 102 min. For the slow component, the plasma half-life varied from 42 to 112 min in the laboratory animals studied; in man, the value was considerably longer with a range of 9.5 to 21 hr (Lin and Symchowicz, 1975).

Urinary excretion of griseofulvin following an oral dose was essentially complete after 48 hr, but was dependent on dosage form. In man, for example, the values ranged from 24% to 59% when the dose was a micronized tablet or a dispersed solid (Chiou and Riegelman, 1971). A species difference can be seen when man, mouse, and rat are compared after receiving orally an aqueous suspension of the fungicide. Rats excreted 32% of the dose, while man and the mouse excreted 38 and 60% of the dose after 48 hr. After an intravenous injection, the rabbit excreted 78% of the dose in urine after 24 hr, while the mouse, man, rat, and dog excreted 70, 58, 53, and 52% of the dose within 48 hr after being dosed. Following oral administration of [$^{14}$C]-griseofulvin as a suspension, 18 and 36% of the doses were excreted in the feces of the mouse and man after 96 hr (Lin and Symchowicz, 1975).

Symchowicz et al. (1967) have demonstrated that biliary excretion of griseofulvin plays an important role in the rat and a lesser role in the rabbit. In the rat griseofulvin metabolism was characterized by extensive biliary excretion and enterohepatic circulation, whereas, excretion of griseofulvin equivalents *via* the rabbit bile represented a minor excretory pathway. In bile duct-cannulated rats during a 24-hr period, about 77% of an intravenous dose of [$^{14}$C]-griseofulvin was found in the bile and 12% in the urine, whereas in the rabbit only 11% was observed in bile and 78% in the urine. The major metabolite in rat bile was 4-demethylgriseofulvin (Fig. 11, II), mainly as the conjugated derivative (Fig. 11, V) with small amounts of the 6-demethylgriseofulvin (Fig. 11, III) as the nonconjugated and conjugated metabolite (Fig. 11, IV) in about equal amounts. In rabbit bile, the major metabolite was 6-demethylgriseofulvin (about equally as conjugated and nonconjugated material) with very small amounts of conjugated 4-demethylgriseofulvin.

The predominant urinary metabolite excreted by rats and rabbits after receiving an intraveneous dose of griseofulvin is 6-demethylgriseofulvin. However, it is very apparent that demethylation at the 6-position is the prefered pathway of metabolism in the rabbit since this metabolite accounted for 95% of the radioactive material extracted from the urine. The comparable value for rat urine was 52% of the radioactivity present (Lin and Symchowicz, 1975). Harris and Riegelman (1969) reported that dogs excreted 40 to 57% of an intravenous dose of griseofulvin in the urine as free 6-demethylgriseofulvin and about 25% of the dose as its glucuronide conjugate. The major metabolite detected in human urine also was 6-demethylgriseofulvin. This metabolite accounted for 84% of the total radioactivity in the urine and was present in about equal amounts as the free and conjugated form. The metabolite, 4-demethylgriseofulvin, was excreted in the urine primarily as the glucuronide and accounted for only about 2% of the radioactivity present in the urine (Lin et al., 1972a).

Lin et al. (1972b) have demonstrated in mice that both 4-demethylgriseofulvin and 6-demethylgriseofulvin are major urinary metabolites; 6-demethylgriseofulvin appeared essentially in the nonconjugated form while 4-demethylgriseofulvin appeared largely in the conjugated form. In general, it may be concluded that in rabbits, dogs, and man the prefered metabolic pathway for metabolism of griseofulvin is via demethylation at the 6-position, while in mice and rats demethylation at the 4-position plays an important role in the metabolism of the fungicide. A significant proportion of the urinary metabolites of orally dosed mice (43%), rabbits (22%), man (12%), and rats (10%) were unidentified metabolites. The dog was unique in that after an intravenous dose of the fungicide all of the radioactivity excreted in the urine was identified as either free or conjugated 6-demethylgriseofulvin.

DISCUSSION

While one could continue to discuss the metabolism of individual fungicides in animals, most of those compounds not included herein have been studied only in the rat, or at least published reports deal only with rats. Nevertheless, those compounds which have been considered serve to point out the types of metabolic reactions common to most fungicides, and allows some comparative aspects relative to animal species to be evaluated.

Although many differences were noted for the individual compounds, the ultimate fate (elimination from the animal body) of the fungicides was often quite similar. In general, fungicides are rapidly absorbed from the gastrointestinal tract, metabolized and excreted largely via the urine. However, the dog often appeared to be an exception where as much as 80% of certain fungicides were excreted in the feces. Again, it should be pointed out that the relatively high levels of fungicide equivalents in the feces of the dog, or in a few cases other animals as well, may result from biliary excretion and not from the lack of absorption of the fungicide. Thus, studies dealing with the bioavailability of fungicides in different species, especially the dog, would seem to warrant further investigation. It might be noted that fungicides are not unique in this respect, since extensive bioavailability studies with most insecticides and herbicides have not been performed.

Species variations in primary or phase I metabolism (e.g., oxidation, reduction, and hydrolysis) of fungicides are for the most part quantitative rather than qualitative in nature. While the same is generally true for secondary or phase II (e.g., conjugation) reactions, these metabolic processes usually show a greater species variation in the metabolism of fungicides.

In many instances the techniques used to identify fungicide metabolites in the different species failed to account for all of the fungicide equivalents appearing in the excreta. For example, the chemical method used to identify thiabendazole and its metabolites in the urine of rats, man, and dogs failed to account for 26, 61, and 77% of the fungicide equivalents appearing therein, thus indicating that the fungicide may have been metabolized by other metabolic pathways. Therefore, the possibility that qualitative differences in the metabolism of thiabendazole and other fungicides considered in this review may exist cannot be ruled out until the unidentified metabolites have been further characterized.

From the information available it may be concluded that, as a class, the organic fungicides are metabolized and excreted in a manner similar to pesticides having insecticidal and/or herbicidal action. While there is an obvious need to more fully determine the metabolic characteristics of fungicides in animals and to assess their significance, this need exists also for most other pesticides. There is no indication that fungicidal activity, *per se*, results in a compound being more or less predictable insofar as metabolite significance in animals is concerned.

ACKNOWLEDGEMENT

The preparation of this review was supported in part by Regional Research Project S-73.

REFERENCES

Anonymous, *Clin. Pharmacol. Therap.* 10, 607 (1969).
Betts, J. J., James, S. P., Thorpe, W. V., *Biochem. J.* 61, 611 (1955).
Borzelleca, J. F., Larson, P. S., Crawford, E. M., Henigar, G. R., Jr., Kuchar, E. J., Klein, H. H., *Toxicol. Appl. Pharmacol.* 18, 522 (1971).

Borzsonyi, M., Sipos, V., Csik, M., *Magy. Onkol.* 19, 175 (1975) (in Hungarian); through *Pestic. Abstr.* 9, 76-0765 (1976).
Bratt, H., Daniel, J. W., Monk, I. H., *Food Cosmet. Toxicol.* 10, 489 (1972).
Briggs, D. E., Waring, R. H., Hackett, A. M., *Pestic. Sci.* 5, 599 (1974).
Calderbank, A., *Acta Phytopathol.* 6, 355 (1971).
Carrigan, P. J., Bates, T. R., *J. Pharm. Sci.* 62, 1476 (1973).
Chin, W. T., Stone, G. M., Smith, A. E., in "International Symposium on Pesticide Terminal Residues", Tohori, A. S., ed., Butterworths, England, 1971, p. 271.
Chiou, W. L., Riegelman, S., *J. Pharm. Sci.* 62, 1376 (1971).
Courtney, D., *Toxicol. Appl. Pharmacol.* 25, 454 (1973).
Cutting, W. C., "Cutting's Handbook of Pharmacology", 4th ed., Appleton-Century-Crofts, New York, 1969, p. 81, 131.
Daniel, J. W., *Biochem. J.* 111, 695 (1969).
Delp, C. J., Klopping, H. L., *Pl. Dis. Reptr.* 52, 95 (1968).
deVos, R. H., Tennoener De Braun, M. C., Olthof, P. D. A., *Bull. Environ. Contam. Toxicol.* 11, 567 (1974).
Douch, P. G. C., *Xenobiotica* 3, 367 (1973).
Douch, P. G. C., *Xenobiotica* 4, 457 (1974).
"Farm Chemicals Handbook", Meister Publishing Co., Willoughby, OH, 1976.
Fleeker, J. R., Lacy, H. H., Schultz, I. R., Houkom, E. C., *J. Agric. Food Chem.* 22, 592 (1974).
Frank, A., *Acta Pharmacol. Toxicol.* 29, 1 (1971).
Gandolfi, A. J., Buhler, D. R., *Xenobiotica* 11, 693 (1974).
Gandolfi, A. J., Dost, F. N., Buhler, D. R., *Fedn. Proc., Fedn. Am. Socs. Exp. Biol.* 31, 605 (1972).
Gardiner, J. A., Brantly, R. K., Sherman, H., *J. Agric. Food Chem.* 16, 1050 (1968).
Gardiner, J. A., Kirkland, J. J., Klopping, H. L., Sherman, H., *J. Agric. Food Chem.* 22, 419 (1974).
Goodman, L. S., Gilman, A., "The Pharmacological Basis of Therapeutics", 4th ed., The MacMillan Company, London, 1970, p. 1302.
Harris, P. A., Riegelman, S., *J. Pharm. Sci.* 58, 93 (1969).
Hashimoto, Y., Makita, T., Nishibe, T., Mori, T., Ohnuma, N. Noguchi, T., Fukuda, Y., *Oyo Yakuri* 6, 873 (1972b) (in Japanese); through CA 78,93337P (1973).
Hashimoto, Y., Makita, T., Ohnuma, N., Noguchi, T., *Toxicol. Appl. Pharmacol.* 23, 606 (1972a).
Hastie, A. C., *Nature* 226, 771 (1970).
Innes, J. R. M., Valerio, M., Ulland, B. M., Pallott, A. J., Petrucelli, L., Fishbein, L., Hart, E. R., Falk, H. L., Klein, M., Peters, A. J., *J. Natl. Cancer Inst.* 42, 1104 (1969).

Jordan, R. L., Borzelleca, J. F., *Toxicol. Appl. Pharmacol.* 25, 454 (1973).
Kappas, A., Georgopouos, S. B., Hastie, A. C., *Mutat. Res.* 26, 17 (1974).
Kennedy, G. L., Jr., Dressler, I. A., Richter, W. R., Keplinger, M. L., and Calandra, J. C., *Toxicol. Appl. Pharmacol.* 35, 137 (1976).
Kimbrough, R. D., *Essays in Toxicol.* 7, 99 (1976).
Kuchar, E. J., Geenty, F. O., Griffith, W. P., Thomas, R. J., *J. Agric. Food Chem.* 17, 1237 (1969).
Leroux, P., Gredt, M., *Phytiat.-Phytopharm.* 21, 45 (1972).
Lin, C., Chang, R., Magat, J., Symchowicz, S., *J. Pharm. Pharmacol.* 24, 911 (1972b).
Lin, C., Magat, J., Chang, R., McGlotten, J., Symchowicz, S., *J. Pharmacol. Exp. Therap.* 187, 415 (1972a).
Lin, C., Symchowicz, S., *Drug Metab. Rev.* 4, 75 (1975).
Lukens, R. J., "Chemistry of Fungicidal Action", Springer-Verlag, New York, 1971.
Lyr, H., Ritter, G., Banasiak, L., *Allg. Mikrobiol.* 14, 313 (1974).
Martin, H., "Pesticide Manual", 3rd Ed., British Crop Protection Council, Droitwich, Eng., 1972.
McManus, E. C., Washko, F. V., Tocco, D. J., *Am. J. Vet. Res.* 27, 849 (1966).
Millar, P. M., *Phytopathology* 59, 1040 (1969).
Mitin, V., Delak, M., Tranger, M., *Vet. Arh.* 39, 318 (1969).
Moll, H. M., Vila, M. C., Heras, C., *Rev. Acad. Cienc.* 28, 373 (1973) (in Spanish); through *Pestic. Abstr.* 7, 74-2682 (1974).
Moreland, D. E., *Ann. Rev. Plant Physiol.* 18, 365 (1967).
Noguchi, T., Ohkuma, K., Kosaka, S., in "International Symposium on Pesticide Terminal Residues", Tohori, A. S., ed., Butterworths, London, 1971, p. 257.
Ohkawa, H., Hisada, Y., Fujiwara, N., Miyamoto, J., *Agric. Biol. Chem.* 38, 1359 (1974).
The Pesticide Review 1975, USDA, Agricultural Stabilization and Conservation Service, Washington, D. C. Prepared by D. L. Fowler and J. N. Mahan, July 1976.
Robinson, H. J., Stoerk, H. E., Graessle, O. E., *Toxicol. Appl. Pharmacol.* 7, 53 (1965).
Rosenblum, C., Tocco, D. J., Howe, E. E., in "Proc. 2nd Ann. Oak Ridge Radioisotope Conf. Gatlinburg, Tenn." 1964, p. 58.
Rowland, M., Riegelman, S., Epstein, W. L., *J. Pharm. Sci.* 57, 984 (1968).
St. John, L. E., Jr. Ammering, J. W., Wagner, D. G., Warner, R. G., Lisk, D. J., *J. Dairy Sci.* 48, 502 (1965).
St. John, L. E., Jr., Lisk, D. J., *J. Agr. Food Chem.* 20, 389 (1972).

Savitt, L. E., *Arch. Dermatol.* 105, 926 (1972).
Seiler, J. P., *Experientia* 29, 622 (1973).
Sherman, H., Culik, R., Jackson, R. A., *Toxicol. Appl. Pharmacol.* 32, 305 (1975).
Smith, R. L., *Ann. Nutr. Alim.* 28, 335 (1974).
Soeda, Y., Kosaka, S., Noguchi, T., *Agric. Biol. Chem.* 36, 817 (1972).
Stefan, A., Stancioiu, N., Ceausescu, S., *An. Univ. Bucuresti, Biol. Anim.* 22, 125 (1973) (in Rumanian); through *CA81*, *131359U* (1974).
Styles, J. A., Garner, R., *Mutat. Res.* 26, 177 (1974).
Symchowicz, S., Staub, M. S., Wong, K. K., *Biochem. Pharmacol.* 16, 2405 (1967).
Thomson, W. T., in "Agricultural Chemicals - Book I Insecticides", Thomson, W. T., ed., Thomson Publications, Indianapolis, Ind., 1973b.
Thomson, W. T., in "Agricultural Chemicals - Book II Herbicides", Thomson, W. T., ed., Thomson Publications, Indianapolis, Ind., 1973a.
Thomson, W. T., in "Agricultural Chemicals - Book IV Fungicides", Thomson, W. T., ed., Thomson Publications, Indianapolis, Ind., 1975.
Tocco, D. J., Buhs, R. P., Brown, H. D., Matzuk, A. R., Mertel, H. E., Harman, R. E., Trenner, N. R., *J. Med. Chem.* 7, 399 (1964).
Tocco, D. J., Egerton, J. R., Bowers, W., Christensen, V. W., Rosenblum, C., *J. Pharmacol. Exptl. Therap.* 149, 263 (1965).
Tocco, D. J., Rosenblum, C., Martin, M., Robinson, H. J., *Toxicol. Appl. Pharmacol.* 9, 31 (1966).
Torchinskiy, A. M., *Vop. Pitan.*, 3, 76 (1973) (in Russian); through *Pestic. Abstr.* 7, 74-2948 (1974).
Ulsamer, A. G., Marzulli, F. N., Coen, R. W., *Food Cosmet. Toxicol.* 11, 625 (1973).
Vogel, E., Chandler, J. L. R., *Experientia* 30, 621 (1974).
Wallnoefer, P. R., Koeniger, M., Safe, S., Hutzinger, O., *Int. J. Environ. Anal. Chem.* 2, 37 (1972).
Waring, R. H., *Xenobiotica* 3, 65 (1973).
Williams, R. T., in "Fundamentals of Drug Metabolism and Drug Disposition", La Du et al., eds., The Williams & Wilkins Co., Baltimore, Md., 1971, p. 187.
Wit, J. G., Van Genderen, H., *Acta Physiol. Pharmacol. Neerl.* 11, 123 (1962).
Woodcock, D., in "Systemic Fungicides", Marsh, R. W., ed., John Wiley and Sons, New York, 1972.

# THE KINETICS OF HALOGENATED HYDROCARBON RETENTION AND ELIMINATION IN DAIRY CATTLE

G. F. Fries

*Pesticide Degradation Laboratory*
*Agricultural Research Service*
*U. S. Department of Agriculture*

*ABSTRACT. A series of similar experiments with chlorinated hydrocarbon pesticides and related industrial compounds are summarized in the context of a model with two open compartments with an irreversible exit to the exterior. Maximum steady state milk excretion was about 0.5 x daily intake with hexachlorobenzene and DDE. The value decreased both with increasing halogenation or ease of metabolism. The more persistent compounds are eliminated with half-lives of 40 to 70 days. The ratio of milk fat level to body fat level ranges from 0.4:1 for polybrominated biphenyls to 1:1 for dieldrin. These measures provide the basic information required to make regulatory and economic decisions in contaminating incidents.*

INTRODUCTION

The persistence and fat solubility of chlorinated hydrocarbon pesticides make this class of compounds one of the more serious causes of residue problems in animals and animal food products. In addition to the chlorinated hydrocarbon pesticides, the chemically related industrial compounds have caused serious contamination problems. These compounds include the polychlorinated biphenyls, polybrominated biphenyls, and hexachlorobenzene.

Most of the chlorinated hydrocarbon pesticides are being phased out; however, they persist in areas of previous usage and the related industrial compounds are still being made and used. Thus, it is desirable to have an understanding of the kinetics of these residues in farm animals.

Two measures are of particular importance in dealing with animal contamination. These are the rates of transfer from diet to animal product under various conditions of intake and the rates of decontamination of animals when they are placed in a clean environment. This information is needed for making both regulatory and farm management decisions. There are many similarities in the behaviors of halogenated hydrocarbons in animals. However, it can be reasonably inferred that the above measures may differ depending on the characteristics of the compounds and the animal.

We have carried out a number of studies with chlorinated hydrocarbon pesticides and related industrial compounds in lactating cows under reasonably standard conditions. This paper is a preliminary summary of our findings within the context of a simple model. The purpose is to derive some general principles that can be applied in future contaminating incidents with either different compounds or animals widely differing in nutritional and physiological status.

DATA BASE

Chemical and common names and abbreviations are shown in Table 1. This paper is confined to our studies, generally

TABLE 1

*Chemical Names of Compounds*

| | |
|---|---|
| DDD | $p,p'$-DDD: 2,2-bis($p$-chlorophenyl)-1,1-dichloroethane |
| | $o,p'$-DDD: 2-($o$-chlorophenyl)-2($p$-chlorophenyl)-1,1-dichloroethane |
| | $p,p'$-Isomer when not specified. |
| DDE | 1,1-Dichloro-2,2-bis($p$-chlorophenyl)ethylene |
| DDT | $p,p'$-DDT: 1,1,1-trichloro-2,2-bis($p$-chlorophenyl)-ethane |
| | $o,p'$-DDT: 1,1,1-trichloro-2-($o$-chlorophenyl)-2-($p$-chlorophenyl)ethane |
| | $p,p'$-Isomer when not specified. |
| Dieldrin | 1,2,3,4,10,10-Hexachloro-6,7-epoxy-1,4,4a,5,6,7,8,8a-octahydro-1,4-*endo-exo*-5,8-dimethanonaphthalene |
| HCB | Hexachlorobenzene |
| PBB | The major hexabromobiphenyl component of Firemaster BP-6 (Michigan Chemical Co., St. Louis, MO). |
| PCB | The chlorobiphenyls occurring as residues from feeding Aroclor 1254 (Monsanto Chemical Co., St. Louis, MO) |

published, to provide uniformity in data collection methods and to allow access to the details of the original data. The major characteristics of the studies considered are indicated in Table 2. Details of the experiments can be found in the

TABLE 2

Sources of data used

|  | Compounds | Cows (no.) | Amount (mg/day) | Duration (days) |
|---|---|---|---|---|
| Fries et al. 1973 | PCB | 9 | 200 | 60 |
| Fries et al. 1969a | DDE | 3 | 25 | 60 |
|  | DDD | 3 | 25 | 60 |
|  | $p,p'$-DDT | 3 | 25 | 60 |
| Fries et al. 1970 | Dieldrin | 12 | 25 | 20 |
| Fries et al. 1971b | DDE |  | 10 | 20 |
|  | $p,p'$-DDT |  | 50 | 20 |
| Fries et al. 1971a | $p,p'$-DDT | 4 | 100 | 20 |
|  | $o,p'$-DDT | 4 | 100 | 20 |
| Fries and Marrow, 1976 | HCB | 2 | 25 | 60 |
|  | DDE |  | 25 | 60 |
|  | HCB | 3 | 5 | 60 |
|  | DDE |  | 5 | 60 |
| Fries and Marrow, 1975 | PBB | 4 | 10 | 60 |

cited references. Compounds were fed at a constant rate for 20 or 60 days in the concentrate portion of the animals' diet. They were added in an acetone solution that was allowed to evaporate before feeding. In most cases the animals used were in midlactation and the level of milk fat production did not change greatly during the course of observation. Milk fat samples were usually obtained at 5-day intervals during the feeding period and subsequent 60 days. Body fat samples were usually obtained at 30-day intervals.

MODEL

In this summary we are using a simple model consisting of two open compartments with an irreversible exit to the exterior from the first compartment. This is illustrated by

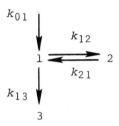

One may visualize compartment 1 as small with a rapid turnover (i.e., blood), and compartment 2 as large with a slower turnover (i.e., body fat). The third compartment shown is equivalent to milk and is a sink, not a true compartment. This model does not take into account elimination by metabolism or enterohepatic recirculation. The net effect of these routes of elimination would be to reduce the value ascribed to input. It would not change the interpretation of the remaining rate constants and their magnitude.

The equations describing the kinetics of this system have been derived and presented in detail by Rescigno and Segre (1966). During the period when the compounds are fed, the amounts in the respective compartments are

$$\frac{X_1}{k_{01}} = \frac{k_{21}}{ab} + \frac{1-k_{21}/a}{b-a} e^{-at} + \frac{1-k_{21}/b}{a-b} e^{-bt} \qquad (1)$$

$$\frac{X_2}{k_{01}} = \frac{k_{12}}{ab} - \frac{k_{12}}{a(b-a)} e^{-at} - \frac{k_{12}}{b(a-b)} e^{-bt} \qquad (2)$$

where $X_1$ and $X_2$ are the amounts of compound in compartments 1 and 2 at time $t$; $e$ is the base of the natural logarithms; $k_{12}$, $k_{21}$ and $k_{13}$ are the transfer coefficients shown above, and

$$a = \tfrac{1}{2}(k_{12} + k_{21} + K_{13}) - [(k_{12} + k_{21} + k_{13})^2 - 4 k_{21} k_{13}]^{\tfrac{1}{2}}) \qquad (3)$$

$$b = \tfrac{1}{2}(k_{12} + k_{21} + k_{13}) + [(K_{12} + k_{21} + k_{13})^2 - 4 k_{21} k_{13}]^{\tfrac{1}{2}}) \qquad (4)$$

Note that $ab = k_{21} k_{13}$.

The amount of compound excreted in milk on day $t$ is

$$\frac{X_3}{k_{01}} = \frac{k_{13}X_1}{k_{01}} = 1 + \frac{k_{13}}{b-a}\, \frac{-b}{}\, e^{-at} + \frac{k_{13}}{a-b}\, \frac{-a}{}\, e^{-bt} \quad (5)$$

If the input of the system occurs in a continuous manner ($k_{01}$) until time $\theta$ and then ceases abruptly, the equations are

$$\frac{X_1}{k_{01}} = -\frac{1-k_{21}/a}{b-a}\, e^{-a(t-\theta)}(1-e^{-a\theta}) - \frac{1-k_{21}/b}{a-b}\, e^{-b(t-\theta)}(1-e^{-a\theta}) \quad (6)$$

$$\frac{X_2}{k_{01}} = \frac{k_{12}}{a(b-a)}\, e^{-a(t-\theta)}(1-e^{-a\theta}) + \frac{k_{12}}{b(a-b)}\, e^{-b(t-\theta)}(1-e^{-b\theta}) \quad (7)$$

$$\frac{X_3}{k_{01}} = -\frac{k_{13}}{b-a}\,\frac{-b}{}\, e^{-a(t-\theta)}(1-e^{-a\theta}) - \frac{k_{13}}{a-b}\,\frac{-a}{}\, e^{-b(t-\theta)}(1-e^{-b\theta}) \quad (8)$$

Any model includes assumptions and simplifications that should be born in mind when interpreting the results. As noted above, we have not considered enterohepatic recirculation nor metabolism in this model. Thus our term $k_{01}$ should be considered the net transfer of compound from the gastrointestinal tract to the circulatory system less metabolism.

The model is derived on the basis of absolute amounts of compounds transferred. The same model would also apply to concentrations of compound in the various compartments but in both cases it is assumed that pool sizes do not change. That is, there is no gain or loss in body fat and milk production remains constant. In reality neither condition is met completely. Our use of animals in midlactation, however, did minimize these changes.

In theory, evaluation of the constants $a$ and $b$ would provide all of the information required to derive the transfer coefficients and to provide complete numerical solutions of the equations. In practice, difficulties are encountered in evaluating $a$ by classical methods because its value is large. The concentrations change rapidly and only a few observations are available even with daily sampling. Conceivably, estimates could be made by using computer analysis that would allow use of all of the data including that during the time the animals were fed the compound. We have not performed this operation to determine if it is practical.

Therefore, we are confining this paper to discussion of a few practical measurements that can be reliably estimated. When considered within the framework of the model one can make inferences concerning the various aspects of residue elimination.

RESULTS

Representative curves for milk fat concentrations of HCB and DDE during the feed and decontamination periods are presented in Figs. 1 and 2. The characteristics of these curves are similar to those of the other compounds and are consistent with the model by visual inspection.

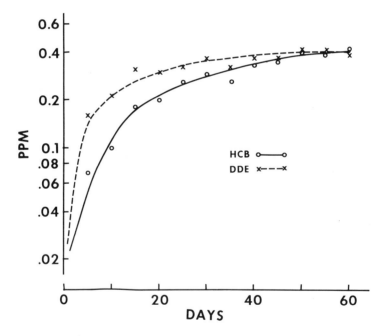

Fig. 1. Average concentrations of HCB and DDE in milk fat while compounds were fed. Each point is an average of six cows and the values were normalized to a 1 mg/day intake to facilitate averaging (From Fries and Marrow, 1976).

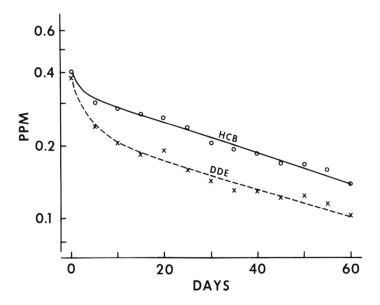

Fig. 2. Average concentrations of HCB and DDE in milk fat of cows after feeding stopped. Each point is an average of five cows and the values were normalized to a 1 mg/day intake to facilitate averaging (From Fries and Marrow, 1976).

The value of $b$ in equation (8) can be readily estimated. As shown in Fig. 2, the first component of equation (8) becomes negligible within 10 to 15 days after feeding stops. Thus, good estimates of $b$ can be made by a least squares fit of the observed milk values to the logarithimic form of the second component of equation (8).

A second measure of practical importance is the concentrations in the compartments that will occur at steady state. This value for milk fat can be estimated by several different methods all of which are based to some degree on the same original information. At steady state $(t=\infty)$, equation (5) will reduce to

$$\frac{X_3}{k_{01}} = \frac{k_{21} k_{13}}{ab} = 1$$

Thus, daily output at steady state is equal to daily input. The steady state value would numerically equal the total output in milk from $t=o$ to $t=\infty$ divided by days fed. If this is divided by the daily feeding rate, one derives a steady state milk concentration normalized to a 1 mg/day intake. Errors of integration are small because most of the output occurred when actual measurements were made. The other methods do not utilize all of the observed values on milk excretion and therefore can be assumed to be less reliable. They are not considered in this paper.

A third measure of great practical and theoretical significance is the relationship between the concentrations in milk fat and body fat. After animals have been on clean feed for at least 15 days, the first terms of equations (7) and (8) are essentially 0 and can be ignored. Thus, dividing the second term of equation (8) by the second term of equation (7) gives the constant

$$\frac{b\ (k_{13}-a)}{k_{12}}$$

The values of the three measures are summarized for the various compounds in Table 3. The amount of compound excreted into milk per day at steady state can be interpreted several ways. The values in Table 3 are the fractions of total intake excreted in milk and are equivalent to the amounts that would be excreted with a 1 mg/day intake. In all cases the recovery from milk is considerably less than the amount of compound fed ranging from a high of .53 mg/day for HCB to a low of .017 mg/day for $o,p'$-DDT.

Values of $b$ are generally consistent among the compounds DDE, PCB, HCB, PBB. There is a possible exception with one short-term experiment with DDE. The values for dieldrin and DDD are significantly greater and average more than twice as great as the above compounds. There is a remarkable consistency for DDD whether it was fed directly or the value is derived from experiments in which $p,p'$-DDT was fed. Residues of $o,p'$-DDT and metabolites were eliminated so rapidly that constants for these could not be evaluated. The value for $p,p'$-DDT is interesting in that it agrees remarkably well with DDE for this measure while the estimated steady state concentration was much lower.

The ratio of milk fat concentration to body fat concentration provides one of the most characteristic and least variable measures. Two of the compounds (HCB and dieldrin)

TABLE 3

*Average steady state rates and levels of excretion, rates of decontamination and ratios of milk fat:body fat levels during decontamination.*

| Compound | Cows | Fed | | Excretion | | | MF:BF |
|---|---|---|---|---|---|---|---|
| | | Amount | Duration | Amount | Level | $b$ | |
| | (no.) | (mg/day) | (days) | (mg/day) | (mg/kg) | (day$^{-1}$) | |
| PCB | 9 | 200 | 60 | .36 $\pm$ .05[1] | .54 $\pm$ .11 | .010 $\pm$ .004 | .64 $\pm$ .06 |
| DDE | 12 | 10 | 20 | .50 $\pm$ .08 | .60 $\pm$ .09 | .021 $\pm$ .006 | .84 $\pm$ .06 |
| | 3 | 5 | 60 | .47 $\pm$ .05 | .70 $\pm$ .09 | .012 $\pm$ .003 | .72 $\pm$ .07 |
| | 2 | 25 | 60 | .35 $\pm$ .01 | .58 $\pm$ .22 | .014 $\pm$ .004 | .81 $\pm$ .02 |
| | 3 | 25 | 60 | .38 $\pm$ .03 | .40 $\pm$ .06 | .012 $\pm$ .001 | .72 $\pm$ .08 |
| HCB | 3 | 5 | 60 | .53 $\pm$ .05 | .80 $\pm$ .10 | .013 $\pm$ .004 | .88 $\pm$ .05 |
| | 2 | 25 | 60 | .34 $\pm$ .05 | .56 $\pm$ .18 | .016 $\pm$ .006 | .93 $\pm$ .06 |
| PBB | 4 | 10 | 60 | .26 $\pm$ .03 | .44 $\pm$ .07 | .012 $\pm$ .030 | .43 $\pm$ .09 |
| Dieldrin | 12 | 25 | 20 | .31 $\pm$ .06 | .37 $\pm$ .04 | .032 $\pm$ .010 | .97 $\pm$ .08 |
| DDD | 3 | 25 | 60 | .07 $\pm$ .01 | .07 $\pm$ .01 | .032 $\pm$ .004 | .16 $\pm$ .02 |
| $p,p'$-DDT DDD | 12 | 50 | 20 | .017 $\pm$ .007 | .014 $\pm$ .007 | N.D.[2] | N.D. |

TABLE 3  continued

| Compound | Cows | Fed Amount | Fed Duration | Excretion Amount | Excretion Level | $b$ | MF:BF |
|---|---|---|---|---|---|---|---|
| | (no.) | (mg/day) | (days) | (mg/day) | (mg/kg) | (day$^{-1}$) | |
| DDT | | | | .034 + .008 | .030 + .008 | .019 + .006 | .62 + .06 |
| Total | | | | .051 + .013 | .044 + .015 | N.D. | N.D. |
| $p,p'$-DDT | 4 | 100 | 20 | | | | |
| DDD | | | | .076 + .013 | .063 + .016 | .065 + .008 | N.D. |
| DDT | | | | .038 + .003 | .032 + .005 | .016 + .001 | N.D. |
| Total | | | | .114 + .015 | .095 + .021 | N.D. | N.D. |
| $p,p'$-DDT | 3 | 25 | 60 | | | | |
| DDD | | | | .024 + .005 | .025 + .006 | .017 + .010 | .18 + .02 |
| DDT | | | | .030 + .003 | .031 + .005 | .010 + .002 | .63 + .09 |
| Total | | | | .055 + .002 | .056 + .008 | N.D. | N.D. |
| $o,p'$-DDT | 4 | 100 | 20 | | | | |
| $o,p'$-DDD | | | | .014 + .006 | .019 + .006 | N.D. | N.D. |
| $o,p'$-DDT | | | | .003 + .001 | .004 + .001 | N.D. | N.D. |
| Total | | | | .017 + .007 | .022 + .007 | N.D. | N.D. |

[1]Standard deviation.
[2]Not determined.

approach a 1:1 ratio. The value for DDD is dramatically lower than any other compound. Among the other compounds the ratio becomes less with increasing molecular weight.

DISCUSSION

The first four compounds listed in Table 2 (PCB, DDE, HCB and PBB) are at most, only sparingly metabolized. This conclusion is based not only on the lack of significant metabolism in other species (Mathews and Anderson, 1975; Mehendale et al., 1975) but also on the failure of these compounds to respond to liver microsomal enzyme induction in cattle (Fries, 1972; Fries et al., 1971b; Hembry et al., 1975; Detering et al., 1975). The remaining compounds are subject to metabolism but the significance of metabolism in eliminating dieldrin may be closely related to whether or not the animal is induced.

If it is assumed that PCB, DDE, HCB and PBB are not metabolized, it follows that the amount recovered in milk is equal to the net transfer of compound from the gastrointestinal tract to the animal body. The net transfer of less than 50% of these compounds suggests that either enterohepatic recirculation is significant or that the absorption of these compounds in cattle under our conditions is considerably less than would often be inferred from small animal experiments. While we have not measured fecal residues in these studies, several reports suggest that this is a very limited means of residue elimination during the decontamination period in ruminants (Detering et al., 1975; Avrahami and Steele, 1972a).

Studies with other species often suggest that absorption of these compounds is rapid and nearly complete (Albro and Fishbein, 1972; Smith et al., 1976; Fries and Marrow, 1975). The apparent difference between cattle and other species can be attributed either to differences in the nature of the digestive tract or to differences in the form that the compound is presented to the animal. The latter point could be of considerable practical importance.

Our studies were conducted with compounds in acetone solution that was allowed to evaporate before feeding. It is possible that the results would have been different if the compounds were fed in oil solution or as an "aged" residue. Marked differences in the form of presentation to the animal have occurred in contamination incidents. For example, the dieldrin incident in Mississippi involved residues in oil, the PBB incident in Michigan involved crystalline material mixed with feed, and the HCB incident in Louisiana involved soil and pasture contamination.

There is some indication that the total dose fed, either as a result of duration or rate will influence the amount excreted into milk. In all cases the highest intake provided the lowest apparent steady state value. Whether this is a true finding or incidental to small numbers cannot be determined. Extensive work with several species suggests that the dose to accumulation ratios of HCB were constant over a wide range of intakes (Avrahami and Gernert, 1972; Avrahami and Steele, 1972abc). In contrast with the few cows in our study there was a higher excretion with the lower rates of intake. This point is worth further investigation because most of our studies were conducted at levels of intake well above those required to produce residues greater than tolerances.

There is evidence that higher molecular weight compounds are less efficiently transferred into the animal's body than those of the lower molecular weights. This is not only true comparing PBB with the other compounds (Table 3) but is a conclusion that can be drawn from other observations. The heptabromobiphenyl component of the PBB mixture had a steady state value of .015 $\pm$ .004 mg/day. This is much lower than the hexabromobiphenyl component. Similar conclusions can be drawn from examination of residue levels of individual PCBs (Smith et al., 1976).

Dieldrin and DDE simultaneously fed to the same cows provides a means of evaluating the importance of metabolism in the elimination of dieldrin. Correlations of steady state excretion ($r = 0.89$) and $b$ ($r = 0.90$) were highly significant ($P < .01$). This indicates that the unmetabolized parent compounds are handled in a similar manner. Recovery of dieldrin was .62 x recovery of DDE indicating that about 40% of the absorbed dieldrin was metabolized. This may not be an uncomplicated evaluation because there is a possibility of microsomal enzyme induction by simultaneously feeding two such compounds.

The amount of DDT and its metabolite, DDD, excreted in the milk was small indicating a great loss by metabolism. However, both the rate constant $b$ and the milk fat to body fat ratio are approximately the same for DDT as DDE. This suggests that, in the cow, the major route of DDT metabolism is conversion to DDD in the rumen (Fries et al., 1969b) with subsequent metabolism of DDD by the animal. The DDT that escaped conversion in the rumen appears to be handled in the same manner as DDE.

As noted, the ratio of milk fat to body fat concentration is a constant and a function of the various individual transfer coefficients. With HCB and dieldrin this ratio approaches 1:1 indicating that the net transfer from the body fat to blood is quite rapid and that milk fat production is the limiting factor in eliminating these compounds. As the transfer coefficients become less, one would expect a lower ratio.

In our work this ratio decreases with increasing molecular weight and becomes least in the case of PBB. The exception, DDD, is readily metabolized and its elimination from the animal is limited by the rate of transfer from fat to blood.

The constancy of the ratio provides suggestions for strategies in dealing with animal decontamination. The rate constant $b$, the level of milk fat production, and the amount of fat in the animal body will be intimately related to the ratio. Increasing milk fat production while maintaining the same body composition should increase the fractional clearance of a compound from body fat and thereby increase the rate constant.

On the other hand, if the amount of body fat is changed without markedly changing milk fat production, several things can happen. Laying down additional fat would reduce milk fat concentration because of dilution but would simultaneously reduce the rate constant because a given level of milk production will clear a smaller fraction of the body fat.

The dilution effect in growing or fattening nonlactating animals is the most important means of reducing residue levels. If an animal is losing weight rapidly, as after calving, it is conceivable that residue levels in milk will actually increase because the concentrating effect of the weight loss is greater than the effect of elimination through milk. A dramatic illustration of this was noted in some of the cows that received massive doses of PBB (Jackson and Halbert, 1974). The cows were quite thin before calving so that fat loss after calving was large relative to the amount of fat in the body at calving. In some cases residue levels more than doubled within 60 to 90 days (Detering et al., 1975). Later, the decline was rapid because each unit of milk fat cleared a large fraction of the animal's remaining body fat. Over a 168 day period the average rate of decline was comparable to our work with normal cows.

The numerous combinations of changes in milk production and body composition make selection of the optimum combination difficult. Computer simulation with an adequate model should be helpful in selecting the optimum conditions. It is possible that the optimum combination would not be nutritionally or physiologically feasible. However, some feasible combinations should be more desirable than others.

CONCLUSIONS AND APPLICATIONS

A typical cow will consume approximately 15 kg of dry matter per day. Thus, a 1 mg/day intake is equivalent to approximately .07 mg/kg in the total diet. The highest projected residue in our studies was 0.8. This suggests that guidelines for feed should be no greater than .10 x guildeline for the milk fat.

It is desirable for regulatory work to have a single tolerance for feed for all species. Investigations are needed with nonlactating animals to determine if steady state levels are similar to those for the lactating animals. Most studies with nonlactating animals are complicated by growth or fattening which is difficult to evaluate.

The elimination rates of the persistent compounds have half-lives in the range of 40 to 70 days. This information is useful in making economic decisions concerning salvage or destruction of contaminated animals. Some field observations suggest that this rate may not continue over long periods of time particularly as residue levels approach the low levels of the tolerances. This could suggest a third compartment with a slow turnover rate that is masked at the higher residue levels usually studied. On the other hand, this could be an artifact caused by continuing low level environmental contamination or introduction of new animals to the herd.

Sampling milk is easier and often more reliable than sampling body fat. The constancy of the milk to body fat ratio makes this a useful means of selecting animals that have body fat levels above guidelines. Since guidelines are usually set at the same level for both body fat and milk fat, it is possible that a cow can produce milk below the guideline and have a body fat level above the guideline. Sampling milk and applying the ratios is a simple and reliable method for identifying these cows.

It is often desirable to know the human exposure in contaminating incidents in order to evaluate the risks associated with the exposure. In the PBB incident maximum exposure occurred six months before identification of the problem. The only way to estimate human exposure was by reconstruction of the milk concentration curve from known times and levels of cow exposure. From this it was estimated that the most highly exposed farm families could have received as much as 10 g PBB from milk alone.

REFERENCES

Albro, P. W., Fishbein, L., *Bull. Environ. Contamin. Toxicol.* 8, 26 (1972).
Avrahami, M., Gernert, I. L., *N. Z. J. Agric. Res.* 15, 783 (1972).
Avrahami, M., Steele, R. T., *N. Z. J. Agric. Res.* 15, 476 (1972a).
Avrahami, M., Steele, R. T., *N. Z. J. Agric. Res.* 15, 482 (1972b).

Avrahami, M., Steele, R. T., *N. Z. J. Agric. Res.* 15, 489 (1972c).
Detering, C. N., Prewitt, L. R., Cook, R. M., Fries, G. F., *J. Animal Sci.* 41, 265 (1975).
Fries, G. F., *Environ. Health Perspect.* 1, 55 (1972).
Fries, G. F., Marrow, G. S., *J. Dairy Sci.* 58, 947 (1975).
Fries, G. F., Marrow, G. S., *J. Dairy Sci.* 59, 475 (1976).
Fries, G. F., Marrow, G. S., Detering, C. N., Prewitt, L. R., Cook, R. M., *J. Dairy Sci.* 58, 764 (1975).
Fries, G. F., Marrow, G. S., Gordon, C. H., *J. Dairy Sci.* 52, 1800 (1969a).
Fries, G. F., Marrow, G. S., Gordon, C. H., *J. Agric. Food Chem.* 17, 860 (1969b).
Fries, G. F., Marrow, G. S., Gordon, C. H., *J. Dairy Sci.* 54, 1870 (1971a).
Fries, G. F., Marrow, G. S., Gordon, C. H., *J. Agric. Food Chem.* 21, 117 (1973).
Fries, G. F., Marrow, G. S., Gordon, C. H., Dryden, L. P., Hartman, A. M., *J. Dairy Sci.* 53, 1632 (1970).
Fries, G. F., Marrow, G. S., Lester, J. W., and Gordon, C. H., *J. Dairy Sci.* 54, 364 (1971b).
Hembry, F. G., Smart, L. I., Binder, T. D., Dixon, J. M., *J. Animal Sci.* 41, 269 (1975).
Jackson, T. F., Halbert, F. L., *J. Amer. Vet. Med. Ass.* 165, 437 (1974).
Mathews, H. B., Anderson, M. W., *Drug Metab. Disp.* 3, 211 (1975).
Mehendale, H. M., Fields, M., Mathews, H. B., *J. Agric. Food Chem.* 23, 261 (1975).
Rescigno, A., Segre, G., "Drug and Tracer Kinetics," Blaisdell Pub. Co., Waltham, Mass., pp. 27-34, 84-87 (1966).
Smith, L. W., Fries, G. F., Weinland, B. T., *J. Dairy Sci.* 59, 465 (1976).

# THE METABOLISM OF $p,p'$-DDT AND $p,p'$-DDE IN THE PIG

G. Sundström, O. Hutzinger,
Laboratory of Environmental Chemistry,
University of Amsterdam

S. Safe and N. Platonow
Department of Chemistry and Biomedical Sciences,
University of Guelph

ABSTRACT. *The urinary metabolites of $p,p'$-DDT and $p,p'$-DDE in the pig have been investigated with emphasis on the detection of phenolic compounds. Upon dosing with $p,p'$-DDT, $p,p'$-DDA was found as the major metabolite. Hydroxylated $p,p'$-DDT or $p,p'$-DDA was not detected, but a minor amount of 3-hydroxy-$p,p'$-DDE was excreted. This phenol was also excreted by pigs given $p,p'$-DDE.*

INTRODUCTION

The metabolism and biological properties of $p,p'$-DDT, its degradation products $p,p'$-DDD and $p,p'$-DDE, and the $o,p'$-analogues of these compounds have been the subject of a large number of investigations (for references see Metcalf, 1973; Fishbein, 1974). In the $p,p'$-series of these compounds metabolic changes have been shown to occur in the aliphatic part of the molecule but in the case of $p,p'$-DDE we recently showed that metabolic alterations also may take place in the aromatic rings. Thus, in environmental samples, excreta from seal and guillemot, we found two phenolic metabolites, isolated as the methyl ethers I and II depicted in Table 1 (Jansson et al., 1975). When rats were fed $p,p'$-DDE, four phenols were found in the feces, and after methylation they were characterized as the methyl ethers I-IV (Table 1) by chromatographic and spectral comparison with synthetic material (Sundström et al., 1975; Sundström, 1976). These findings show that chlorine substitution in the 4-positions of the phenyl rings do not render the aromatic part of the molecule

TABLE 1

Structures and chromatographic properties of compounds investigated

| Compound | $R_F$ value ×100 [a] solvent A | solvent B | Retention time |
|---|---|---|---|
| I  (2-methoxy-2',4-dichloro-DDE-like) | 64 | 54 | 8.9 |
| II (4-methoxy-3,4'-dichloro-DDE-like) | 62 | 44 | 9.7 |
| III (3-methoxy-2,4'-dichloro-DDE-like) | 65 | 54 | 7.3 |
| IV (4-methoxy-4'-chloro-DDE-like) | 64 | 50 | 6.6 |
| V (methyl 4,4'-dichlorodiphenylacetate) | 59 | 40 | 5.8 |

[a] on 0.5 mm layers

unsusceptible to metabolism as has been assumed previously.

The $o,p'$-isomers of DDT and DDD, on the other hand, are extensively metabolized both in the 2-chlorosubstituted phenyl ring and in the aliphatic part of the molecule, and a diversity of metabolites have been identified after feeding of $o,p'$-DDT to rats (Feil et al., 1973) and chickens (Feil et al., 1975) and $o,p'$-DDD to humans (Reif et al., 1974) and rats (Reif and Sinsheimer, 1975).

In our studies on the fate of chlorinated hydrocarbons in domestic animals, we have previously investigated the metabolism of chlorobiphenyls in cows (Safe et al., 1975a), pigs (Safe et al., 1975b) and goats (Safe et al., 1975a; Safe et al., 1975c). The present paper deals with the metabolism of $p,p'$-DDT and $p,p'$-DDE in the pig with special emphasis on the possible formation of phenolic metabolites.

EXPERIMENTAL

*Administration of Compounds*

An amount of $p,p'$-DDT or $p,p'$-DDE (>99% purity, Aldrich Chemical Co.) equal to a dose level of 100 mg/kg was dissolved in corn oil to give a 20% solution. To this solution was added an aqueous phase which consisted of 0.3% Pluronic F68 (BASF, Wyandotte), 0.3% Tween 80 (Atlas Chemicals) and 4% dextrose to give a 10% emulsion of oil and a resultant 2% emulsion of halogen compound. The mixture was sonicated until the particle size was less than 1 micron. The emulsion was administrated by intraperitoneal injection into a four week old female pig housed in a metabolic cage. Commercial feed and water was given *ad libitum* for 7 days during which urine and feces were continually collected and stored at 0°.

*Extraction and Analysis*

The urine was acidified to pH 5 with acetic acid and extracted with diethyl ether to give an extract of unconjugated metabolites. Thereafter the urine was diluted with concentrated sulfuric acid to give a 6 N solution which was heated at 100° for 1 hr. After cooling, the solution was diluted with water and extracted with diethyl ether in order to obtain an extract of metabolites originally present in conjugated form. The dried ($MgSO_4$) extracts were methylated with an excess of potassium carbonate-methyl iodide in acetone solution and finally concentrated.

Fecal samples were extracted with tetrahydrofuran, concentrated, and the residue partitioned between 4N hydrochloric acid and hexane-diethyl ether (1:1). After drying and evaporation of the organic layer, the oily residue was dissolved in hexane and treated with concentrated sulfuric acid. The hexane solution was finally analyzed by combined GLC-mass spectrometry.

*Thin-layer Chromatography*

The concentrated urine extracts were initially purified on 2 mm precoated silica gel plates (Kieselgel 60 $F_{254}$, Merck) which were developed with chloroform (solvent A) and subsequently on 0.5 mm layers of the same adsorbent using hexane-ethyl acetate (4:1, solvent B) as eluent. Synthetic methyl ethers of previously identified phenolic metabolites (Sundström, 1976) were used as standards in the chromatographic work as was the methyl ester of $p,p'$-DDA (Table 1).

*GLC-Mass Spectrometry*

Final analysis of extracts was performed by combined GLC-mass spectrometry using a Hewlett-Packard 5982A GLC-mass spectrometer system operating in the electron-impact mode. The column (290 X 0.2 cm) was packed with 0.2% Carbowax 20M on 100-120 mesh Chromosorb W (Aue et al., 1973). The gas flow (He) was about 25 ml/min and the oven was programmed from 180° (2 min) to 220° at 4°/min. Injection temperature was 250° and the mass spectra were obtained at 70 eV.

Final characterization of metabolites was done by comparison of retention times and mass spectra of authentic compounds (Sundström et al., 1975; Sundström, 1976).

RESULTS AND DISCUSSION

The methylated extracts of urinary metabolites were fractionated by TLC before investigation by GLC-mass spectrometry. Although the majority of the phenolic metabolites of $p,p'$-DDE, $o,p'$-DDT and $o,p'$-DDD excreted by the rat were found in the feces (Sundström, 1976; Feil et al., 1973; Reif and Sinsheimer, 1975) it was not possible to use the fecal material in the search for polar metabolites from the pig due to the presence of very large amounts of co-extractives. Only the occurrence in the feces of unpolar metabolites of $p,p'$-DDT was therefore investigated.

Selected ion chromatograms were obtained from the fractions eluted from the thin-layers using molecular ions or fragment ions characteristic for the different types of compounds expected to be present in the samples. In the case of the methyl ethers of phenolic metabolites of $p,p'$-DDE (Table 1) the intense molecular ions at 346 (compounds I-III) and 312 (compound IV) mass units (m.u.) were chosen. In the mass spectra of DDT compounds with saturated aliphatic parts like $p,p$-DDT, $p,p$-DDD and $p,p$-DDA, including its methyl ester, the fragment at 235 m.u. constitutes the base peak. This fragment ion is also expected to form the base peak in spectra of the methyl ethers of phenolic metabolites of this type of compounds. In addition to these types of metabolites, the presence of compounds originally containing two phenolic groups, or one phenol and one methoxy group, was also investigated.

*Urinary Metabolites of $p,p'$-DDE*

The GLC-mass spectral investigation of the derivatized urine extract from pigs fed $p,p'$-DDE revealed the presence of the methylated phenol I (Table 1 and Fig. 1). Most of this phenol was excreted in the free form and only traces could be detected in the extract of the hydrolyzed urine. In the case of the rat, the methyl ethers I-IV were isolated in a ratio of 100:4:8:1.2, respectively, (Sundström, 1976) and the GLC and mass spectrometric conditions used in this study should

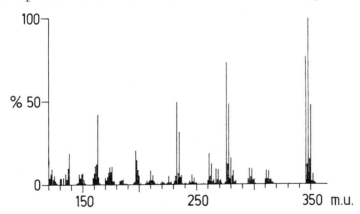

Fig. 1. Partial mass spectrum (EI) of compound I isolated from pig urine.

readily allow the identification of these compounds if present in these proportions. However, although there were some indications of the presence of the methyl ethers II and III, the amounts were too small to obtain mass spectra for positive characterization. This may indicate that the metabolic conversion in the pig of p,p'-DDE to hydroxylated metabolites proceeds by routes different from those in the rat. It is worth mentioning that p,p'-DDA was not found in the urine of pigs fed p,p'-DDE.

## Urinary Metabolites of p,p'-DDT

Analysis of the urine from pigs fed p,p'-DDT showed that p,p'-DDA is the major urinary metabolite. p,p'-DDA was present both free and conjugated although the nature of the conjugate, or conjugates, was not investigated. In the extract of the unhydrolyzed urine, small but significant amounts of compound I, the methyl ether of the major urinary metabolite of p,p'-DDE, were present (Fig. 2). We were not able to detect any phenolic compounds in which the aliphatic part of the p,p'-DDT molecule was unaffected. Nor did we find any indications of phenolic derivatives of p,p'-DDA.

These findings suggest that after oxidation of p,p'-DDT to p,p'-DDA, the latter compound is rapidly excreted and is not further accessible to enzymatic action. Likewise, enzymatic hydroxylation does not occur in the unreactive 4-chlorophenyl rings as long as the more reactive aliphatic part of the p,p'-DDT molecule is still intact. This behavior is in contrast to that of the o,p'-isomers of DDT and DDD in which both the 2-chlorophenyl ring and the aliphatic parts are involved in enzymatic oxidation reactions (Feil et al., 1973; Feil et al., 1975; Reif et al., 1974; Reif and Sinsheimer, 1975).

Thus, the metabolic conversion of p,p'-DDT in the pig seems to follow two main routes. One route involves the oxidation of the aliphatic part of the molecule to p,p'-DDA which is excreted, free or conjugated, without further conversion. In the other route p,p'-DDT is converted to p,p'-DDE, which is then excreted in the urine as such or after enzymatic hydroxylation. p,p'-DDE was the only compound found in the feces of pigs administered p,p'-DDT

The findings that DDT compounds are metabolized to phenolic derivatives raises questions as to whether these metabolites are responsible for any biological effects observed after feeding the DDT compounds to animals. It was suggested by Welch et al. (1969) that phenolic metabolites of p,p'-DDT were the ultimate cause for the estrogenic activity of this compound. The similarity in structure of some metabolites of o,p'-DDT to some highly estrogenic compounds (Bitman and Cecil, 1970; Grundy, 1957) was recently pointed out by Feil et al. (1973) who indicated that these metabolites may be responsible for the effects of o,p'-DDT on reproduction in several species. Likewise, it was suggested that the biotransformation of o,p'-DDD to a 4-hydroxylated species is responsible for its adrenolytic activity (Reif et al., 1974).

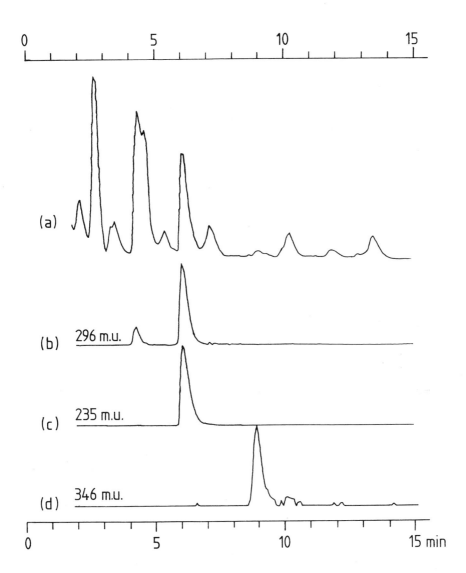

Fig. 2. Ion chromatograms of the methylated extract from unhydrolyzed urine of pigs fed p,p'-DDT. (a) Total ion chromatogram; (b) ion chromatogram displaying the molecular ion of p,p'-DDA methyl ester; (c) the base peak in the mass spectrum of this compound and (d) the molecular ion of the methylated p,p'-DDE metabolites I-III.

ACKNOWLEDGEMENT

A grant from the North Atlantic Treaty Organization is gratefully acknowledged.

REFERENCES

Aue, W. A., Hastings, C. R., Kapila, S., *J. Chromatog.* 77, 299 (1973).
Bitman, J., Cecil, H. C., *J. Agric. Food Chem.* 18, 1108 (1970).
Feil, V. J., Lamoureux, C. H., Styrvoky, E., Zaylskie, R. G., Thacker, E. J., Holman, G. M., *J. Agric. Food Chem.* 21, 1072 (1973).
Feil, V. J., Lamoureux, C. H., Zaylskie, R. G., *J. Agric. Food Chem.* 23, 382 (1975).
Fishbein, L., *J. Chromatog.* 98, 177 (1974).
Grundy, J., *Chem. Rev.* 57, 281 (1957).
Jansson, B., Jensen, S., Olsson, M., Renberg, L., Sundström, G., Vaz, R., *Ambio* 4, 93 (1975).
Metcalf, R. L., *J. Agric. Food Chem.* 21, 511 (1973).
Reif, V. D., Sinsheimer, J. E., Ward, J. C., Schteingart, D. E., *J. Pharm. Sci.* 63, 1730 (1974).
Reif, V. D., Sinsheimer, J. E., *Drug Metab. Disp.* 3, 15 (1975).
Safe, S., Platonow, N., Hutzinger, O., *J. Agric. Food Chem.* 23, 259 (1975a).
Safe, S., Ruzo, L. O., Jones, D., Platonow, N. S., Hutzinger, O., *Can. J. Physiol. Pharmacol.* 53, 392 (1975b).
Safe, S., Platonow, N., Hutzinger, O., Jamieson, W. D., *Biomed. Mass Spectrom.* 2, 201 (1975c).
Sundström, G., Jansson, B., Jensen, S., *Nature (London)* 255, 627 (1975).
Sundström, G., *J. Agric. Food Chem.*, submitted for publication (1976).
Welch, R. M., Levin, W., Conney, A. H., *Toxicol. Appl. Pharmacol.* 14, 358 (1969).

# MIREX, CHLORDANE, DIELDRIN, DDT, AND PCB'S: METABOLITES AND PHOTOISOMERS IN L. ONTARIO HERRING GULLS

D. J. Hallett, R. J. Norstrom,
*Toxic Chemicals Division*
*Canadian Wildlife Service*

F. I. Onuska and M. Comba
*Analytical Methods Development Research Section*
*Canada Centre for Inland Waters*

ABSTRACT. When the adult body lipid and eggs of herring gulls (Larus argentatus) breeding on L. Ontario were analyzed by gas chromatography using an EC detector, high residue levels of PCB's, DDE, HCB, and dieldrin were found. Extensive column chromatographic separation using cocoanut charcoal to remove high PCB interference coupled with a long, narrowbore, Florisil column allowed the identification by GLC/MS of mirex, photomirex, β-BHC, cis-chlordane, trans-nonachlor, photo cis-nonachlor, oxychlordane, heptachlor epoxide, p,p'-DDD, and methoxychlor. Tri- and tetrachlorobiphenyls accounted for 32 percent of the total PCB's accumulated in adult lipid. Analysis of saponified extracts revealed the presence of hydroxy metabolites of tetra-, penta, hexa-, and heptachlorobiphenyls in the adult bird lipid. Herring gull eggs contained the same organochlorine residues and hydroxy metabolites of PCB's found in the adult tissues, indicating transfer from female to egg.

The herring gull (*Larus argentatus*) is a long-lived predator with opportunistic feeding habits. Being relatively non-migratory in L. Ontario and having a position at the top of an aquatic food chain makes it an excellent species for indicating the environmental health of the whole lake ecosystem.
In 1975 there was almost total reproductive failure of herring gulls in L. Ontario (Gilman et al., 1976). A study of eggs that failed to hatch indicated that the major problems on L. Ontario were the disappearance of eggs and the failure of

embryos to develop. The failure of embryos to develop may be due in part to embryotoxic substances transmitted from the female (Gilman et al., 1976). A search for persistent pollutants with potential embryotoxicity was therefore undertaken using GLC and GLC/MS.

Table 1 shows industrial organochlorine contaminants found in the adult body lipid of Great Lakes herring gulls.

TABLE 1

*Industrial pollutants identified in Lake Ontario herring gull lipid*

| Compound | Concentration in lipid (ppm) |
|---|---|
| PCB's (Aroclor 1260) | 3530 |
| Mirex | 220 |
| Photomirex | 84 |
| HCB | 6.7 |

PCB's were calculated to be 3530 mg/kg as Arochlor 1260 or as decachlorobiphenyl after perchlorination. Monohydro mirex was named photomirex since it has been shown to form by photodegradation (Alley and Layton, 1974; Gibson et al., 1972) and not biodegradation in aquatic ecosystems (Metcalf et al., 1973). We have synthesized photomirex by the method of Kecher et al., (1974) and the value of 84 mg/kg shown in Table 1 was obtained using an analytical standard of the compound. Mirex and photomirex were determined using the Hall electrolytic conductivity detector and were confirmed by GLC/MS after perchlorination (Hallett et al., 1976). Photomirex has been shown to bioaccumulate and have a pattern of persistence similar to mirex in rats (Gibson et al., 1972).

The mass spectra of mirex, photomirex, and kepone all show ion abundances at m/e 270 and 235 (Cl = 35), indicative of retro Diels-Alder fragmentation to hexachlorocyclopentadiene (m/e 270), with subsequent loss of chlorine to the m/e 235 fragment. Photomirex also shows a pentachlorocyclopentadiene ion ($C_5HCl_5^+$) at m/e 236. The base peak for photomirex at m/e 201 shows a further Cl loss. Also evident are very weak ion fragments at m/e 508 and 510 with a stronger ion abundance at m/e 471 for the M-35 loss which supported a molecular composition of $C_{10}HCl_{11}$ for monohydro mirex.

Table 2 shows the concentrations of organochlorine pesticides found in the adult body lipid of L. Ontario herring gulls.

TABLE 2

*Pesticide residues identified in Lake Ontario herring gull lipid*

| Compound | Concentration in lipid (ppm) |
|---|---|
| β-BHC | 34.6 |
| p,p'-DDT | 3.4 |
| p,p'-DDE | 310 |
| p,p'-DDD | 1.8 |
| cis-chlordane[1] | <1.0 |
| trans-nonachlor[1] | <1.0 |
| photo cis-nonachlor[1] | <1.0 |
| oxychlordane | 1.0 |
| heptachlor epoxide | 1.0 |
| monohydro oxychlordane | <1.0 |
| dieldrin | 5.6 |
| methoxychlor[1] | <1.0 |

[1] Identified by GLC/MS only

The β isomer of BHC is the most persistent of the three isomers, and was the only one found. Six distinct residues related to technical chlordane were evident and confirmed by GLC/MS. Clean column chromatographic separations of each compound were achieved by first removing PCB's using the cocoanut charcoal technique of Berg et al. (1972), followed by a 1.2 cm I.D. X 72 cm long column of 1.2% water-deactivated Florisil, which was consecutively eluted with 8 X 50 ml hexane followed by 2 X 100 ml of 15% methylene chloride/hexane and 2 X 200 ml of 30% methylene chloride/hexane. *Cis*-chlordane and *trans*-nonachlor (Fig. 1) were identified as residues by retention time and by GLC/MS, heptachlor epoxide (mass spectrum Fig. 2) and oxychlordane (mass spectrum Fig. 3) were found as metabolites (confirmed on EC and Hall detectors), and two compounds were identified by MS alone as photoproducts of chlordane. The mass spectrum in Fig. 4 was identified from the molecular ion at m/e 406, and ion abundances at m/e 371, 270 and 235 as photo *cis*-chlordane, previously reported by

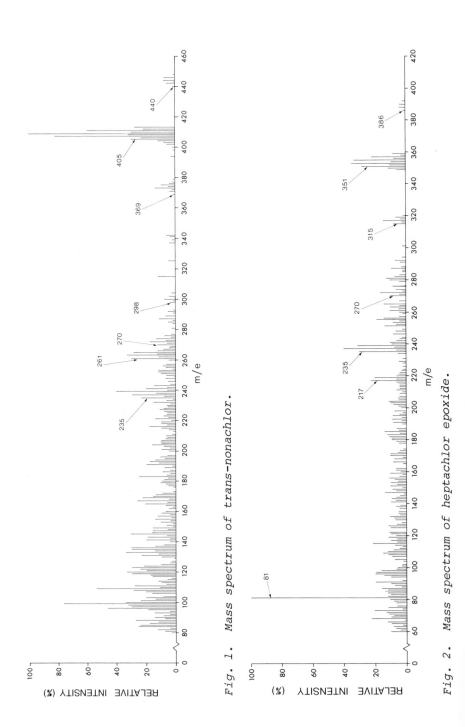

Fig. 1. Mass spectrum of trans-nonachlor.

Fig. 2. Mass spectrum of heptachlor epoxide.

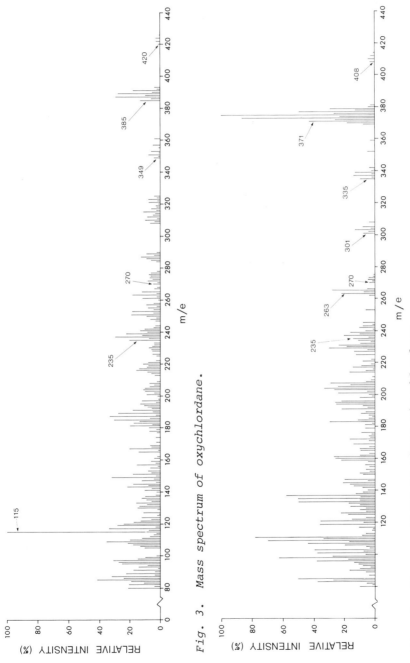

Fig. 3. Mass spectrum of oxychlordane.

Fig. 4. Mass spectrum of photo cis-chlordane.

Onuska et al. (1976). A further chlordane related product was identified by a mass spectrum indicative of monohydro oxychlordane. Residues of dieldrin (5.6 mg/kg) were found but no photodieldrin accumulated in the gulls.

An EC chromatogram of PCB residues showed a complex pattern containing mostly penta-, hexa-, and heptachlorobiphphenyls corresponding to those found in Aroclors 1260 and 1254 (Hallett et al., 1976). For this reason, PCB's were measured as mixtures of Aroclor 1260/1254 (Reynolds, 1972), Webb and McCall, 1973) or by perchlorination (Armour, 1973).

The extract was chromatographed using a GLC/MS/FID system and mass spectra for each peak (Fig. 5) were recorded. Tri-, tetra-, penta-, hexa-, hepta-, and octachlorobiphenyls were identified. The relative response of these homologues on FID is within $\pm$ 10%. The peaks were integrated and the percentage of the total taken. Trichlorobiphenyls represented 4.6%, tetra's 27.4%, penta's 20.5%, hexa's 41%, hepta's 5.5% and octachlorobiphenyls less than 1% of the total chromatogram. The percentage of individual isomers was not determined but progress is now being made using glass capillary columns. The electron capture chromatogram was not representative of the PCB residue since Aroclor 1260 does not contain any tri- or tetrachlorobiphenyls and Aroclor 1254 contains only 11% tetrachlorobiphenyls.

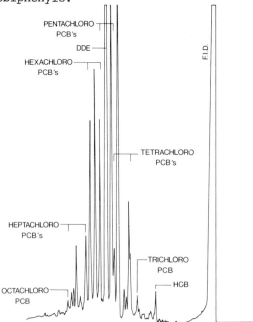

Fig. 5. Chromatogram of herring gull extract. Flame ionization detector, 3% OV-101 column, 100°-260°C at 6°/min.

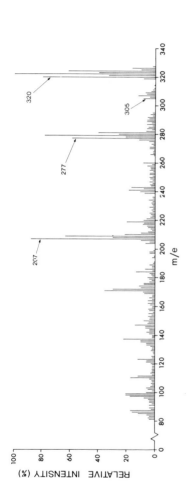

Fig. 6. Mass spectrum of tetrachlorobiphenyl methyl ether.

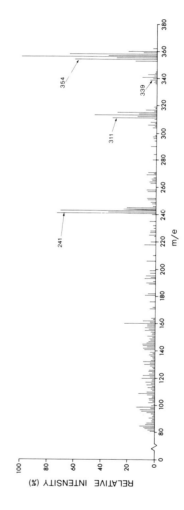

Fig. 7. Mass spectrum of pentachlorobiphenyl methyl ether.

189

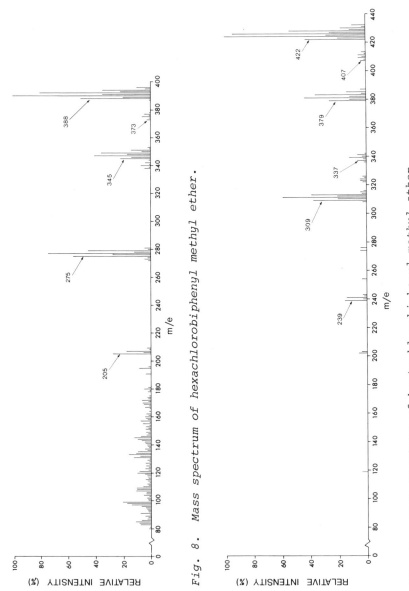

Fig. 8. Mass spectrum of hexachlorobiphenyl methyl ether.

Fig. 9. Mass spectrum of heptachlorobiphenyl methyl ether.

The same lipid was extracted after saponification in methanolic KOH (Grimmer et al., 1975) and was chromatographed as before. All methylene chloride eluates contained a series of electron capturing peaks not seen in unsaponified extracts. The identities of the peaks were determined by GLC/MS to be methoxy-PCB's containing 4,5,6, and 7 chlorines. The mass spectra (Figs. 6,7,8, and 9) show the same strong molecular ion as very toxic chlorinated dibenzo-p-dioxins of similar chlorine substitution. However, the methoxylated PCB's exhibit an M-43 loss (M-COCH$_3$) and not the M-63 loss (M-COCl) diagnostic of a chlorinated dibenzodioxin molecule. Hydroxylated metabolites of polychlorinated biphenyls containing 4,5,6, and 7 chlorines have previously been found in the feces of Baltic guillemots and seals (Janson et al., 1975). These were identified after derivatization with diazomethane to form their corresponding methyl ethers. Safe et al., (1974) have identified some methoxy-PCB's as minor urinary metabolites. We assume that the bulk of the methoxy-PCB's identified here resulted from methylation of hydroxy derivatives during saponification. Aroclor 1254 did not produce methoxylated derivatives during the same saponification procedure. Further extraction, derivatization and GLC/MS studies are underway. Saponified extracts of herring gull eggs from L. Ontario also contained methoxy-tetrachlorobiphenyls, indicating that it is possible for the presumed hydroxy PCB metabolites to be transferred from female to egg.

Residues accumulated by the herring gull demonstrate that considerable photodegradation and metabolic degradation occurred before or during the biotransfer process through the lake ecosystem. These degradation products may be considerably more toxic than the parent residues but are not usually detected in routine monitoring procedures. They should not be overlooked. We are now considering the toxicity of these degradation products as related to the reproductive failure of L. Ontario herring gulls.

REFERENCES

Alley, E. G. and Layton, B. R., "Mass Spectrometry and NMR Spectroscopy in Pesticide Chemistry," Haque, R. and Biros, F. J. (eds.) Plenum Press, New York, NY, pp. 81-90 (1974).
Armour, J. A., J. Assoc. Offic. Anal. Chem. 56, 987 (1973).
Berg, D. W., Diosady, P. L., and Rees, G. A. V., Bull. Environ. Contam. Toxicol. 7, 338 (1972).
Gibson, J. R., Ivie, G. W., and Dorough, H. W., J. Agric. Food Chem., 1246 (1972).

Gilman, A. P., Fox, G. A., Peakall, D. B., Teeple, S. M., Carroll, T. R., and Haymes, G., *J. Wildl. Manage.*, in press (1976).

Grimmer, G., and Bohnke, H., *J. Assoc. Offic. Anal. Chem.* 58, 725 (1975).

Hallett, D. J., Norstrom, R. J., Onuska, F. I., Comba, M. E. and Sampson, R., *J. Agric. Food Chem.*, in press (1976).

Janson, R., Jensen, S., Olsson, M., Renberg, L., Sundstrom, G. and Vaz, R., *Ambio,* 93 (1975).

Kecher, R. M., Skibinskaya, M. B., Gallai, O. S., and Zetirov, N. S., *Zhurnal Organichoskoi Khimii,* 10, 411 (1974).

Metcalf, R. L., Kapoor, I. P., Lu, P. Y., Schuth, C. K. and Sherman, P., *Environ. Health Perspec. Exptl.* Issue No. 4, 35 (1973).

Onuska, F. I. and Comba, M. E., *Biomedical Mass Spectrometry,* 2, 176 (1975).

Reynolds, L., *Residue Reviews* 34, 27 (1971).

Safe, S., Hutzinger, O., and Jones, D., *J. Agric. Food Chem.,* 23, 851 (1975).

Webb, R. G. and McCall, A. C., *J. Chromatogr. Sci.*, 11, 366 (1973).

DDT METABOLISM IN PENNSYLVANIA WHITE-TAIL DEER

D. A. Kurtz and J. L. George

The Pennsylvania State University

ABSTRACT. *Technical grade DDT was fed to Pennsylvania white-tail deer in five experiments lasting 100 days at several dose levels to determine the chronic and acute effects on the deer as well as the nature and distribution of DDT metabolites in selected tissues and organs, feces, and urine. The dosage of three experiments was constant at levels of $10-10^4$ µg/g DDT in the feed. The dosage of the two remaining experiments was attenuated, beginning at levels of 50, 175, or 350 µg/g and ending at 1 µg/g. Total DDT concentrations found in feces and urine were an order of magnitude below feeding concentrations. Highest concentrations found from constant diets were in bone marrow, fat, and liver tissues; lower amounts in kidney, brain, and muscle; and lowest amounts in blood. For attenuated diets, concentrations found in these tissues were negligible. TDE was the major metabolite found in tissues and organs, ranging from 91-31% ($\bar{X}=56\%$) of the total DDT type residues in liver, kidney, fat, blood, muscle, brain, and bone marrow. TDE in feces was 65% of the total residue. DDA was the major metabolite in urine, comprising 87% of the total residue.*

INTRODUCTION

Because of its wide and effective use, its low cost, and especially because of its persistence, DDT (Table 1) has been studied probably more than any other pesticide. Though no longer permitted to be used generally in the United States, it continues to be studied primarily due to its long environmental half-life. One of the best ways of studying the presence of DDT in wild, uncultured areas is by analysis of wildlife inhabiting these areas. In Pennsylvania the white-tail deer is not only plentiful but also a valuable wildlife and game resource. It should be a good indicator of the presence

TABLE 1

*Chemical names of DDT metabolites*

| Common name | Chemical name |
|---|---|
| DDT | 1,1,1-trichloro-2,2-bis(p-chlorophenyl) ethane |
| DDE | 1,1-dichloro-2,2-bis(p-chlorophenyl) ethylene |
| TDE | 1,1-dichloro-2,2-bis(p-chlorophenyl) ethane |
| DDA | bis(p-chlorophenyl)acetic acid |

of DDT and other persistent chlorinated compounds. It is a minor food source, and is also representative of ruminant large animals. Thus, there is the need for a complete study of the absorption rates, body reservoirs, and depletion routes and rates for DDT in deer. This study provides information in these areas.

Studies of DDT content, metabolism, and depletion involving wildlife or large animals are few in number. The only deer feeding studies have been by Watson et al. (1975), who used only two animals, and by Pillmore and Finley (1963), who used only one animal. All were mule deer, *Odocoileus hemionus*. Wild populations of both mule and white-tail deer, *O. virginianus*, have been studied in DDT treated and control areas. Walker et al. (1965) analyzed 199 deer in Idaho and Washington for DDT content in adipose tissues. Again in Idaho, Benson and Smith (1972) analyzed adipose tissue from 93 mule deer from DDT treated and control areas. Greenwood et al. (1967) analyzed a general collection of animals in South Dakota in 1964 following a long history of aerial DDT spraying. Baetcke et al. (1972) in Mississippi, and Jewel (1967) in Colorado reported similar studies. Deer in Alabama soybean fields were analyzed for DDT residues by Causey et al. (1972).

Other wild ruminants, such as elk, moose, and antelope were studied by Walker et al. (1965), Moore et al. (1968) in South Dakota, Benson et al. (1973) in Idaho, and Greenwood et al. (1967). Benson et al. (1974) also studied bears in Idaho. These studies involved for the most part adipose tissues, with few analyses on other tissues and organs. Finally, for the esoterically minded, reference is made to analyses of DDT residues in Rhodesian elephant and impala liver samples (Billing and Phelps, 1972).

By way of comparison as to dietary intake, fat reservoirs, and depletion rates, beef steers were studied by Wilson et al. (1970, 1971a, 1971b). In these studies steers were fed apple pomace containing trace levels of DDT in several dietary regimens. The residue analyses were on adipose tissue.

The current study involves the feeding of 45 wild Pennsylvania white-tail deer temporarily housed in cages or pens. Two types of feeding schedules at several levels of DDT were followed. Fecal and urine samples were taken where appropriate and analyzed during the course of the 100-day feeding period. At the conclusion of the study period tissue and organ samples were also analyzed.

EXPERIMENTAL

The work was done in five experiments as is summarized in Table 2. Experiments I, II, and III involved addition of DDT at a constantly dosed concentration in the total diet for the

TABLE 2

Experimental setup showing age and number of deer in each experiment and their environment

| Exp. No. | Age of Deer | No. of Deer | Environment |
|---|---|---|---|
| I | Yearling | 11 | Metabolism cage |
| II | Fawn | 10 | Metabolism cage |
| III | Fawn | 7 | Pen |
| IV | Fawn | 7 | Pen |
| V | 1,2,5-year olds | 10 | Pen |

duration of the experiment. DDT was introduced by dosing the standard complete ration feed at the Penn State Deer Research Facility. Sex-paired animals were fed at four dose levels of 10, 100, 1000, and 10,000 µg/g of DDT in feed (Fig. 1). Experiments IV and V involved introduction of DDT at attentuated concentration in a decreasing direction. Control deer and three dose levels of 50, 175, and 350 µg/g in the diet were used, and these were lowered each 10-day period (Fig. 1) until they reached 1 µg/g during the last 10 days.

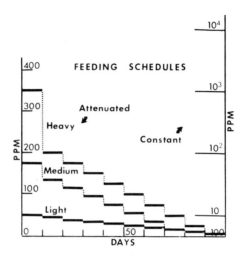

Fig. 1. Scheduled plan for feeding DDT-dosed rations. Constantly dosed rations for Experiments I-III are shown in the right vertical coordinate on a log scale. The three levels of attenuated diet for Experiments IV-V are shown on the left vertical coordinate on an arithmetic scale.

Feeding Procedures

Rations were spiked with acetone solutions of DDT to give the appropriate concentrations, and the dosed feed mixtures were subsequently analyzed. When weighted to the 10,000 µg/g concentration, the standard deviation of these analyses was ± 3526 µg/g.

The treated feed was added to feed boxes in the cages such that feed was always available to the deer. Spilled feed was subtracted from feed added to obtain net feed consumed. Daily records were kept of feed added and feces and urine collected. If feed was not added for one or more days, the feed last added was averaged over skipped days to obtain the daily consumed feed data.

It should be noted that these experiments were conducted during the three winter months, a time when deer are losing weight and drawing from their fat reserves. During this time their feeding is quite irregular. Differences in feed consumption were noted over periods of days. Fig. 2 gives two examples of feeding patterns. A typical example was one where daily averages ranged from 0.75 to 1.36 kg of feed. One of the most extreme examples showed a feeding ranging from 0.13 to 0.92 kg of feed per day. The feeding of 0.13-0.14 lasted over a period of 20 days. These variable rates of feeding could not be predicted. Caution is therefore given in interpreting data in respect to daily feeding variances. Schwartz and Nagy (1974) report also that dry matter digestion of feed dosed with 1000 µg/g DDT was significantly decreased ($P < 0.05$) for both penned and wild deer.

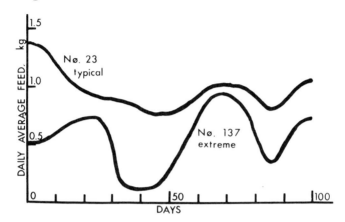

Fig. 2. Examples of typical and extreme feeding patterns of individual deer averaged over 10-day periods.

Complete data on feeding patterns and information on all analytical aspects will be reported elsewhere.

Deer in Experiments I and II were kept in metabolism cages (Cowan, 1969) to enable urine sampling. Those in the remaining experiments were kept in 20 foot square pens. The metabolism cages were very restrictive for animal movement, but the pens allowed some freedom. Nevertheless it has been shown that there is a significant difference ($P < 0.05$) in dry matter digestion of control level feeds between penned and wild deer (Schwartz and Nagy, 1974).

*Collection of Samples*

Feces and urine were sampled at 0, 2, 10, 30, and 100 days from the start of each experiment, and their weights recorded. Table 3 shows two examples of the relative sizes of deer used in this study, amounts of feed consumed, and total weights of feces and urine collected.

Tissues and organs were collected at the conclusion of the 100-day feeding or at death if earlier than 100 days. Bone marrow was sampled from the femur. Fat was sampled from the visceral, subcutaneous, renal, and/or omental areas. The

TABLE 3

Feed, feces, urine, and weight relationships

| Experiment (Deer) | Animal size Start-finish kg | Total feed 100 days kg[1] | Total feces 100 days kg[1] | Total urine 100 days kg[1] |
|---|---|---|---|---|
| I (90) Yearling Female | 53-48 | 54 | 15 | 54 |
| II (18) Fawn Male | 30-26 | 69 | 19 | 73 |

[1]Air dried basis.

muscle sampled was the gastrocnemius. The right hemisphere of the brain or the medulla was analyzed, and kidney samples were taken from the cortex. Blood and liver were sampled generally.

*Analysis*

*Extraction.* Tissues and organ samples were extracted by the micromethod outlined by Thompson (1971). Feces were oven dried and soxhlet extracted with a mixture of hexane and isopropanol. Urine samples were extracted by the method of Cranmer (1969). They were subjected to ethylation with diazoethane (Stanley, 1966).

*Cleanup.* All samples were cleaned up by florisil column chromatography by the method described by Thompson (1971).

*Quantification.* Residues were determined by gas liquid chromatography utilizing several chromatographs and numerous columns. The bulk of the analyses were performed using columns packed either with 3% OV-1 or QF-1/OV-17 (2:1) loaded 0.05% on glass beads (GLC-110). Confirmation work was accomplished on a 3% DEGS column. The gas chromatographs employed were a Research Specialties 600 instrument ($Sr^{90}$ detector), a Packard Model 7935 instrument ($Ni^{63}$ detector) and a Microtek Model 220 instrument ($Ni^{63}$ detector). Quantification was accomplished with either an Infotronics Model CRS-11-HSB or Model CRS-100 area integrator.

Recoveries of DDE, TDE, and DDT were 94, 93, and 97%, respectively, from vegetable oil. Tissue, organ, and feces samples were not corrected. Recoveries from urine for DDE,

TDE, and DDT were 93, 88, and 86%, respectively, and for DDA were 50-92% depending on amount added. Urine samples were corrected for recovery.

RESULTS AND DISCUSSION

The results of the study of total DDT concentrations in various tissues and organs, feces, and urine samples will be discussed first. The metabolic composition of the DDT residues as related to conversion of the parent compound will then be discussed. Feces residues are reported on a dried basis while urine and tissues are reported on a whole sample or wet basis.

*Total DDT Residues: Constant DDT in Feed*

*Tissues and organs.* At the end of the 100 day experiments, all males and some females were sacrificed in order to obtain specimens of various tissues and organs for analysis. Tissues from animals that died before the end of the experiment were likewise collected. Tissues analyzed were fat, bone marrow, liver, kidney, brain, muscle, and blood. Table 4 contains data obtained. The data were grouped according to the total DDT input to each animal: control animals were combined together; inputs from 14 through 31 mg DDT/kg body weight ($\bar{X}$=23) formed the next group; those 112-314 ($\bar{X}$=204) formed the third group; and those 1000-4400 ($\bar{X}$=2500) comprised the final group. The plots for each tissue as the mean DDT consumed/body weight *versus* mean DDT concentration in the tissues are found in Fig. 3.

Fat contained the highest concentrations of DDT. For tissues from animals in the middle grouping (204 mg DDT consumed/kg body weight), fat contained 125 µg/g total DDT. Bone marrow contained approximately half that amount, 53 µg/g. Substantially lower concentrations were found in liver, 4.5; kidney, 2.5; muscle, 1.7; and brain 1.1 µg/g. Blood contained <0.1 µg/g. These data are based on whole, wet tissue.

An interesting application of these data is to project the levels of DDT in the diet of wild deer by examination of the residue data. Such data are listed in Table 5. White-tail and mule deer from South Dakota, Mississippi, Alabama, Colorado, Idaho, and Washington were analyzed by various researchers in general and sprayed areas. Concentrations of total DDT found in these studies are found in the fourth column of the table. In the fifth column are listed the expected DDT concentrations in the diet of these deer. Caution should be given in using these projections since the experimental feeding time in this work was only 100 days while that in the other reports is of variable time periods up to two or more years. In addition

TABLE 4

*Total residues versus DDT consumed in body tissues and organs of Experiments I-III*[1]

| Total DDT intake/ body wt. means mg/kg | No. data pts. | Bone marrow | Total DDT residues | | | | | |
|---|---|---|---|---|---|---|---|---|
| | | | Brain | Fat | Muscle | Blood | Kidney | Liver |
| | | | | µg/g | | | | |
| Controls | 3-5 | 0.7 | <0.1 | 0.8 | <0.1 | <0.1 | <0.1 | 0.1 |
| 23 | 4 | 4.9 | <0.1 | 7.5 | <0.1 | <0.1 | 0.5 | 0.9 |
| 204 | 5 | 53 | 1.1 | 125 | 1.7 | <0.1 | 2.5 | 4.5 |
| 2500 | 7 | 2000 | 59 | | 32 | 1.5 | 56 | 822 |

[1]Data based on wet tissue analysis

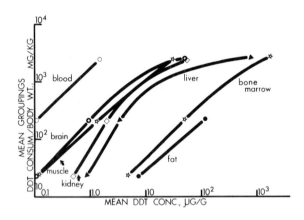

Fig. 3. Total DDT residues versus DDT consumed in body tissues and organs of Experiments I-III.

our study was done in the winter months when feeding was irregular and weight and fat reduction were taking place. Nonetheless some indication as to the local contamination of feed can be found in Table 5.

TABLE 5

Expected concentration of DDT in diet of U. S. deer based on analyses of this work from adipose/fat or muscle tissues

| Deer | State | Identification | Conc. found in fat ppm | Expected DDT conc. in diet mg/kg | Reference |
|---|---|---|---|---|---|
| wt[1] | SD | General | 0.10 | 0.8 | Greenwood et al.(1967) |
| wt | MS | General | 0.29 | 1.8 | Baetcke et al. (1972) |
| wt | SC | Control area | 0.14 | 1.0 | Barrier (1970) |
|  |  | Spray area | 1.76 | 7.5 |  |
| wt | AL | Untreated soybean | 0.35[2] | 65. | Causey et al. (1972) |
|  |  | Treated soybean | 3.0 [2] | 400. |  |
| Mule | MT | Bitterroot Natl. For. 4 mo. post spray | 9. | 27. | Pillmore and Finley (1963) |
|  | MT | Gallatin Natl. For. 2 mos. post spray | 21.5 | 55. |  |
|  | NM | Santa Fe Natl. For. 1-5 mos. post spray | 6.0 | 20. |  |
|  | CO | Rio Grande Natl. For. 4 mo. post spray | 13.1 | 37. |  |
| Mule | CO | General | 2.8 | 11. | Jewel (1967) |
| Mule | ID | Control | 0.08 | 0.7 | Benson and Smith (1972) |
|  |  | Spray 1964 | 20. | 50. |  |
|  |  | 1969 - Control | 0.05 | 0.5 |  |
|  |  | 1969 - Spray '64 area | 0.18 | 1.3 |  |
| [3] | ID & WA | General | 0.12 | 0.9 | Walker et al. (1965) |

[1] wt = white-tail    [2] muscle tissue    [3] Unknown

Female mule deer fed DDT at the rate of 15.2 g over a period of 67 days (Pillmore and Finley, 1963) were found to contain 80 and 2 µg/g total DDT in fat and liver, respectively. These values are about half those obtained in our work.

Another aspect of interest is the feed or food concentrations required to induce a level of one µg/g of total DDT in adipose or fat tissues. Our work has found that this ratio is dependent on the feed level. From other works calculations were made (Table 6) assuming proportionate increase in DDT residues in fat with respect to time over a period of 100 days. Mule deer were fed at a high rate over 67 days and found to require higher amounts of DDT feed to obtain DDT residue concentrations equal to those for Pennsylvania white-tail deer. Cattle were fed at very low amounts for 80-84 days and found to require lower amounts than for white-tail deer. Finally man was treated with moderate amounts over 183 days and required much lower amounts of DDT in the food to induce similar levels in fat. These requirements must certainly be related to the fat content of the respective body studied as well as

TABLE 6

Comparing DDT accumulation in fat of Pennsylvania white-tail deer with other mammals

| Other species | | DDT feeding level over 100 days mg/kg | Accumulation efficiency DDT feed level/DDT in fat mg/kg per µg/g | | PA deer[2] |
|---|---|---|---|---|---|
| | | | | Reference | |
| Mule deer, female | | 453. | 3.80 | Pillmore & Finley (1963) | 1.51 |
| Cattle, | Angus | 1.76[1] | 2.41 | Wilson et al. (1971) | 5.87 |
| | Hereford | 2.32[1] | 1.53 | Wilson et al. (1970) | 5.80 |
| | Holstein | 2.15[1] | 1.46 | Wilson et al. (1970) | 5.81 |
| Man | | 52. | 0.80 | Morgan & Roan (1971) | 2.48 |

[1] Mixed metabolites DDE, TDE, and DDT in feed.
[2] Current study.

the rate of fat deposition when dosed.

The trends found in Fig. 3 will now be examined. If the contamination of tissues and organs is truly proportionate, then each curve on this log-log plot should have a slope of 45°. The curves for both fat and bone marrow are nearly so. Also those for muscle, brain, and blood have a similar slope. However, in examining the liver curve two observations are noted. First, the curve rises steeper than 45°. Second, at the highest DDT consumption level the concentrations in the liver become much *greater* than expected! A similar observation is made for the curve for kidney. It is these two organs that are most deeply involved in detoxification and elimination of foreign compounds in mammals.

The flow of nutritional and toxicant materials in the body of the deer, as in all ruminants, begins after ingestion in the rumen. Some materials are absorbed directly from the rumen and some after passage to the intestines. Those not absorbed, and some that are resecreted into the gastrointestinal tract, eventually appear in feces. Absorbed materials pass into the body proper *via* the liver where some transformations undoubtedly take place. Some products leaving the liver *via* the blood stream are deposited in fat, muscle, and other body tissues, and some pass into the kidney where water soluble toxicants are removed from the body through the urine.

Leng (1976) has suggested that ingested pesticides or toxicants are handled in different ways depending on the total amount of influx. For low concentrations the body may be able to handle the detoxification quite adequately. However, like a dam bursting due to the pressure of the water behind it, larger amounts of the toxicant may not be handled adequately and the material floods the tissues and body. This may be the effect seen here with the kidney and liver. There is also the possibility of seeing this effect in the brain tissues where the blood-brain barrier has been overwhelmed by high concentrations of DDT.

*Feces.* DDT not absorbed through the intestines passes out of the animal through the feces. Data for the analysis of fecal material is presented in Table 7 and Fig. 4. Groupings of DDT ingested were again the same as for tissues and organs. At each consumption level there appears to be a maximum point of DDT concentration between 10 and 30 days and then a lowering by the 100th day. This suggests that the initial stress of the appearance of DDT prevents absorption into the body; later the body becomes acclimatized and absorbs proportionately more DDT. At highest levels DDT in feces was 5000 µg/g at 10 days and 2220 at 30 days only to lower all the way to 109 at 100 days, a reduction factor of up to 50 to 1. At the middle level DDT concentration was found to be 47 µg/g at the 30th day only to be 17 at the 100th day, this time a factor of only 3:1. At the lowest ingested level the factor is 6:1. This data here again suggest some unusual bodily processes were occurring at high DDT concentrations.

TABLE 7

*Total residues versus DDT intake in feces of Experiments I-III*[1]

| Total DDT consumed/ body wt., mean mg/kg | No. of deer | Total DDT residues in µg/g Sampling day | | | | |
|---|---|---|---|---|---|---|
| | | 0 | 2 | 10 | 30 | 100 |
| Controls | 2-5 | 3.2 | 1.8 | 2.7 | 6.7 | 0.1 |
| 23 | 1-5 | 1.2 | 2.5 | 4.9 | 2.9 | 0.8 |
| 204 | 2-4 | 0.7 | 14.8 | 3.8 | 47 | 16.7 |
| 2500 | 3-7 | 0.7 | 224 | 4950 | 2220 | 109 |

[1]Data based on dried material

Watson et al. (1975) studied fecal residues of mule deer fawns fed DDT at the rate of 5 mg per day. This amounts to 0.2 mg/kg/day (assuming 25 kg fawn weight) which was about equivalent to our lowest level. At 30 days they obtained a total DDT concentration of 0.5 µg/g in the feces, which was lower than our finding of about 4 µg/g. This amounts to a higher absorption rate at this level for their studies.

*Urine.* The highest DDT concentrations were in urine found at the 100th day at all feeding levels (Table 8, Fig. 5).

TABLE 8

*Total residues versus DDT intake in urine of Experiments I-II*[1]

| Total DDT consumed/ body wt., mean mg/kg | No. of deer | Total DDT residues in µg/g Sampling day | | | | |
|---|---|---|---|---|---|---|
| | | 0 | 2 | 10 | 30 | 100 |
| Controls | 1-4 | <0.1 | <0.1 | ns[2] | <0.1 | <0.1 |
| 23 | 1-4 | <0.1 | <0.1 | ns | 0.5 | 0.6 |
| 204 | 1-3 | <0.1 | 0.4 | ns | 5.5 | 7.1 |
| 2500 | 3-4 | <0.1 | 12.0 | ns | 41.0 | 80.0 |

[1] Data based on whole urine
[2] ns - no sample

At the feeding level of 23 mg DDT/kg body weight the 30th and 100th day DDT concentrations were 0.5 and 0.6 µg/g. For the 204 mg/kg level they were 5.5 and 7.1 µg/g, respectively. Finally for the highest feeding level they were 41 and 80 µg/g, respectively.

The body and kidney were effective in maintaining the flow of DDT into urine even at the highest ingested levels. The comparison of DDT concentrations with levels in urine indicates a proportionate relationship. However, it is noted that the urine levels are all about 1/10 of those of feces at the 30th day.

*Total DDT Residues: Attenuated DDT in Feed*

*Tissues and organs.* The feeding of decreasing amounts of pesticide in this experiment represents a unique approach, heretofore untried, for DDT. It simulates the ingestion by animals in a situation where a single applied dose to flora will gradually dissipate from the environment. It is quite evident that smaller amounts of residues were present at the conclusion of the studies (Table 9) than for those of constant

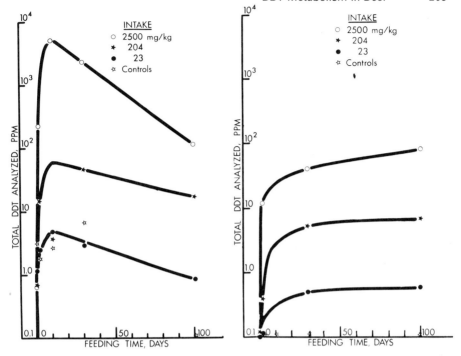

Fig. 4. Total DDT residues versus DDT intake in feces of Experiments I-III.

Fig. 5. Total DDT residues versus DDT intake in urine of Experiments I-II.

feeding studies (Table 4). Unfortunately these depletion studies produced a paucity of data. Most females were reserved for later experiments and not sacrificed at 100 days. Two deer died before the end of the 100-day term. Examination of the data show that the single deer on a diet of 141 mg DDT/kg body weight had consistently less residues in organs than those on either the 60 or 294 mg/kg diet.

The data were grouped into three groups according to the DDT intake per body weight: the first group was 60 mg/kg, the second was 141 mg/kg and the third ranged from 254 to 334 ($\bar{X}=294$) mg/kg. DDT content in tissues and organs will be discussed at the highest bodily intake level of 294 mg DDT/kg body weight (Table 9). Highest concentrations of total DDT in animals on the attenuated diets at the end of 100 days were found in the bone marrow, 11 µg/g. Fat and liver contained 3.1 and 1.9 µg/g, respectively. Brain contained 1.0 µg/g while concentrations in kidney and muscle were both 0.2 µg/g. DDT was not found in the blood at the level of sensitivity tested, 0.1 µg/g.

Precise relationships between the experiments can be

TABLE 9

*Total residues versus DDT consumed in body tissues and organs of Experiments IV-V[1]. Male deer fed 100 days.*

| Total DDT intake/ body wt. means mg/kg | No. data pts. | Total DDT residue | | | | | | |
|---|---|---|---|---|---|---|---|---|
| | | Bone marrow | Brain | Fat | Muscle | Blood | Kidney | Liver |
| | | | | µg/g | | | | |
| Controls | 1-2 | 0.1 | <0.1 | 0.1 | <0.1 | <0.1 | <0.1 | <0.1 |
| 60 | 2 | 6.7 | 0.1 | 5.7 | 0.1 | <0.1 | 0.1 | 1.0 |
| 141 | 1 | 3.7 | <0.1 | ns[2] | 0.1 | <0.1 | <0.1 | 0.6 |
| 294 | 2 | 11.0 | 1.0 | 3.1 | 0.2 | <0.1 | 0.2 | 1.9 |

[1]Data based on wet tissue
[2]ns = no sample

determined by consulting Fig. 3, which shows concentrations of total DDT in tissues and organs as a function of bodily intake for constant feeding experiments. Discussion here will center on the data from an attenuated diet of 294 mg/kg diet. DDT in fat had a very high mobility under the conditions of these experiments, for the fat of deer on the attenuated diet contained 98% less DDT. A high mobility was also observed for DDT in bone marrow (84%). That in kidney and muscle also were very highly mobile in having 90-92% less DDT in the animals on attenuated diets. The DDT in liver on the other hand was only 55% lower, and brain DDT residues were only 30% lower on the attenuated diets than on the constant level diet.

It is doubtful that such high depletion rates would be found for concentrations of DDT below those used in this study. Martin et al. (1976) determined the depletion rate of total DDT in perianal fat of steers starting at a concentration of 10 ppm. After 28 days the rate was insignificant (P <0.5) and leveled off at 56 days to only a 45% loss. Wilson et al. (1971b) found that Angus steers, having 0.70 µg/g DDT in biopsied cod fat, depleted the DDT only to 0.46 µg/g, a 34% loss, after 75 days. On the other hand the nature of deer, its environment, and its general activity could produce higher depletion rates even at low concentrations of DDT than for steers or other domesticated ruminants. No data are available to support this hypothesis, however.

Depletion rate calculations for this type of attenuated

diet is a complex function and will not be discussed here. Pure depletion was studied and the rate constants calculated by Reynolds et al. (1975) for sheep from adipose tissue. Starting from a total DDT level of 630 ppm, depletion was followed for a period of 112 days, and rates for DDT, DDE, and TDE were calculated. Davies et al. (1971) determined the depletion rate for man in a study of 11 subjects. Starting at concentrations of 2.1 and 7.0 ppm for DDT and DDE, respectively, they found the half lives of these two analogs to be 26 and 18 months.

*Feces.* Feces from deer on the attenuated diets were also analyzed for DDT. As expected (Table 10 and Fig. 6) the DDT residues peaked early—about the 10th day—then fell to 1 µg/g (on 307 mg/kg diet) to less than 0.2 µg/g (lower diets at the end of the experiment. The feces from the deer receiving 60 mg DDT/kg body weight were found to contain 14.4 µg/g DDT at the 10th day. Those from deer on diets ranging from 101 to 141 ($\bar{X}$=130) mg DDT/kg body weight contained 48 µg/g DDT after 10 days, while those on diets ranging from 254 to 334 ($\bar{X}$=307) mg DDT/kg peaked at 94 µg/g at the 10th day (Table 10).

TABLE 10

Total residues versus DDT intake in feces of Experiments IV-V[1]

| Total DDT consumed/ body wt., mean mg/kg | No. of deer | Total DDT residues in µg/g Sampling day | | | | |
|---|---|---|---|---|---|---|
| | | 0 | 2 | 10 | 30 | 100 |
| Controls | 3-5 | 3.0 | 0.3 | 0.9 | 0.9 | <0.2 |
| 60 | 3-4 | 0.3 | 1.6 | 14.4 | 8.3 | <0.2 |
| 130 | 4 | 2.8 | 4.6 | 48.0 | 19.7 | <0.2 |
| 307 | 3 | 1.6 | 15.7 | 94.0 | 73.0 | 1.1 |

[1]Data based on dried material

The relationship of the loss of DDT through feces between the constant feeding and attenuated experiments was examined. Since consumption rates of DDT varied between the two types of experiments, the total DDT was found in feces for each of the feeding levels in each of the experiments by integration of DDT concentration under the curves presented by Figs. 4 and 6. DDT concentration was summed for every 10th day. These data are presented in Table 11. When plotted (Fig. 7), the constant

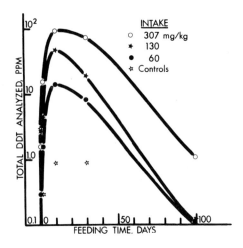

Fig. 6. Total DDT residues versus DDT intake in feces of Experiments IV-V.

Table 11

DDT consumed/body weight versus integrated DDT concentration in feces comparing constant DDT diet (Exp. I-III) versus attenuated DDT diet (Exp. IV-V)

| DDT diet | DDT consumed/ body weight mg/kg | Integrated DDT conc. µg/g |
|---|---|---|
| Constant, Experiments I-III | 23 | 28 |
|  | 204 | 400 |
|  | 2500 | 14,900 |
| Attenuated, Experiments IV-V | 60 | 46 |
|  | 130 | 133 |
|  | 307 | 356 |

diet curve was found to be almost linear, though not proportional. Surprisingly, the curve for the attenuated diet did not superimpose but rather was found above that for the constant diet. The result was that for a given composition of

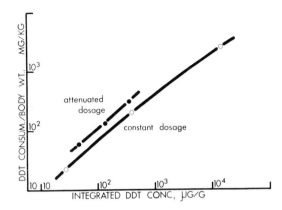

*Fig. 7. Relationship of the DDT consumed/body weight and integrated DDT concentration in feces comparing constant DDT diet (Experiments I-III) versus attenuated DDT diet (Experiments IV-V).*

DDT in diet an attenuated diet produced *less* DDT in feces. This amount was about 56% less for attenuated diets in the range of 60 to 300 mg DDT/kg body weight.

*Conversion of DDT to Metabolic Products*

The DDT analogs included in the study of the fate of DDT in the white-tail deer were DDE, TDE (DDD), and in the case of urine, DDA. A discussion of the relative quantities of these analogs follows.

*Feces.* As food containing DDT is ingested potential metabolism in the alimentary tract can take place in the rumen, in the enterohepatic circulation through the liver, and in the intestines. The total effect of each of these reaction sites is reflected in the analysis of the feces. An 83% conversion from these sources were observed on an overall basis from all five experiments. TDE was found to be the major metabolite and amounted to 65% of the total feces residues. DDE made up the remaining 18% in the metabolites in feces (Table 12). When a much more restrictive sample population was taken from the first three experiments, those of constant feeding, from the 10th, 30th, and 100th day, the relative abundance of these analogs was approximately the same.

TABLE 12

Conversion of DDT in deer in feces, organs and tissues, and urine

| Portion analyzed (Experiments) | No. of data points | DDT analog found | | | | Relative conc. ppm[1] |
|---|---|---|---|---|---|---|
| | | DDE | TDE | DDA | DDT | |
| | | | | % | | |
| Feces (I-III)[2] | 20 | 20 | 66 | na[3] | 14 | |
| (I-V) | 74 | 18 | 65 | na | 17 | |
| Tissues and Organs | | | | | | |
| Overall | 76 | 6 | 56 | na | 38 | |
| Liver | 12 | 7 | 91 | na | 2 | 3.6 |
| Blood | 4 | 12 | 58 | na | 30 | 0.1 |
| Muscle | 8 | 7 | 58 | na | 35 | 1.3 |
| Fat | 10 | 2 | 60 | na | 38 | 125 |
| Bone marrow | 22 | 4 | 31 | na | 65 | 53 |
| Brain | 10 | 7 | 44 | na | 49 | 0.9 |
| Kidney | 10 | 7 | 73 | na | 20 | 2.0 |
| Urine (I-II)[2] | 17 | + | 1.3 | 98 | 0.7 | |
| (I-II) | 34 | 2.6 | 3.4 | 87 | 7 | |

[1] Total residue at consumption level of 204 mg/kg body weight
[2] For 10, 30, and 100 days
[3] na = no analysis

In working with bovine rumen fluid *in vitro*, Sink et al. (1972) found a moderate amount of metabolism of DDT, 7% from a roughage diet and 20% from a concentrate diet. DDE was the predominate metabolite over TDE by a ratio of 4 to 9 times. Should the effects between the two animals be similar, then it is clear that further metabolism took place in other parts of the alimentary canal.

Watson et al. (1975) studied the fate of DDT in a single mule deer fawn. Thirty days after feeding 5 mg per day it was found that there was 90% metabolism, 89% TDE, the remaining 1% being DDE. The TDE content was somewhat higher than found in our study.

*Tissues and organs.* The overall conversion of DDT in tissues and organs was found to be 62% (56% TDE, 6% DDE). A much better picture of the system is seen by examination of the individual organs and tissues analyzed.

Unmetabolized DDT in the liver amounted to 2%. The major metabolite was TDE (91% of the total DDT), and small amounts of DDE (7%) (Table 12). Analysis of other tissue showed much larger amounts of unmetabolized DDT. Unconverted DDT in blood was 30% of the total DDT and in the storage tissues of muscle, fat, and bone marrow was 35, 38, and 65%, respectfully. That in the brain was 49% of the total DDT content. It may be that rapid TDE conversion to DDA (Roan et al., 1971; Reif et al., 1975) effectively increases the proportion of DDT level at the expense of TDE. However, no DDA analyses were undertaken of the tissues and organs. The higher proportion of unmetabolized DDT in bone marrow (65%) as compared to that in fat and muscle may be accounted for lower DDT mobility. In actual concentrations, fat contained 47.5 µg/g DDT while bone marrow had 34.5. TDE contents of these tissues were 75 and 16 µg/g, respectively. Hence fat was a better storage depot for TDE and bone marrow a better depot for DDT.

Blood serving as the major transportation for these analogs contained more DDE, 12%, than any other tissue (maximum of 7%). DDE seemed to store better in muscle or bone marrow than in fat though differences were not great. The blood-brain barrier admitted higher than average percentages of DDT, 49%, and lower TDE, 44%. The kidney appeared to selectively store TDE, as only 20% of the total DDT was found to be DDT whereas 73% was TDE.

*Urine.* DDT residues in the urine were almost totally in the form of DDA (Table 12). For the selected samples from the 10th, 30th, and 100th days 98% of the total DDT was DDA with only 1.3% TDE and 0.7% DDT. For all urine samples conversion products were 87% DDA, 2.6% DDE, and 3.4% TDE for a total of 93% converted products.

Fat analyses from a variety of other animals lead to some interesting comparisons. Pillmore and Finley (1963) found that a female mule deer fed DDT retained residues in fat amounting to 62% DDT, 38% TDE, and 0% DDE which indicated a lower conversion rate by almost half when compared with our study. In national forest areas sprayed with DDT, these authors found an even lower conversion rate where the total residues were from 6 to 13 µg/g total DDT. In the fat of these deer 95% of the total DDT was DDT and only 4% TDE.

Benson and Smith (1972) found even more interesting results in their study of Idaho mule deer. In control areas 62% of the residues in fat were DDT, none as TDE and 38% as DDE while in recently sprayed areas, fat contained 89% as DDT, 8% as TDE, and only 3% DDE. It is possible that control area ingestion may have included large percentages of DDE as environmental degraded DDT. Causey et al. (1972) did not see such an effect in studying white-tail deer in Alabama soybean fields. Percentages of the various analogs varied little between the deer that foraged in treated and untreated areas. DDT was found to be 57 and 54% and TDE was 20 and 14%, respectively. In these samples DDE was high at 23 and 32%, respectively.

Wild moose in Idaho (Benson et al., 1973) were found to contain percentages of the analogs similar to control areas for deer. DDT was found at 57%, TDE at 8% and DDE at 33%.

Whiting et al. (1973) have reported fat analyses resulting from long term feeding of DDT to cows. In these studies, 77 to 82% of the total DDT was DDE, the major metabolite. After feeding sheep, on the other hand, Reynolds et al. (1975) found a much smaller conversion and the major metabolite was TDE. DDT content was 84%, TDE 12% and DDE 4%. In one feeding study of man (Morgan and Roan, 1971) adipose analysis following ingestion of 20 mg per day for 183 days gave products containing 81% DDT, 3% TDE, and 16% DDE. Roan et al. (1971) also studied the storage of DDA and DDE in man following ingestion of these two analogs. They found that DDA was excreted very rapidly with no storage while DDE did not produce any DDA as an excreted product.

*Mortality*

A number of the deer in these experiments died during or shortly after the study period. Data presented in Table 13 clearly show increasing mortality with increasing dosage. At dosage rates of 3 to 3000 mg/kg the mortality was generally about 20%. However, when dosages were above 3000, it was 100%.

TABLE 13

*Mortality*

| DDT intake/body wt. mg/kg | Deaths-Total | Mortality % |
|---|---|---|
| 0-2.9 | 1-11 | 9 |
| 3-29 | 1-5 | 20 |
| 30-299 | 3-15 | 20 |
| 300-2999 | 2-9 | 22 |
| 3000- | 5-5 | 100 |

Fawns born from females in the experiments also gave information in this respect. The data in Table 14 show that fawns born from mothers who had received less than 16 mg DDT/kg body weight had a reasonable survival expectancy. Those born from mothers receiving 100 or more mg DDT/kg body weight, however, had no life expectancy at all as all died in two months or less.

TABLE 14

Survival of fawns from DDT-treated deer

| Dam<br>DDT consumed/<br>body wt.<br>mg/kg | Sire<br>DDT consumed/<br>body wt.<br>mg/kg | Fawn<br>survival time<br><br>days |
|---|---|---|
| 0 | 0 | 100+ |
| 0 | 14.1 | 100+ |
| 0 | 14.1 | 16 |
| 15.9 | 14.1 | 100+ |
| 15.9 | 14.1 | 216 |
| 112 | 14.1 | 5 |
| 112 | 14.1 | 0.5 |
| 101 | 0 | 67 |
| 137 | 0 | 9 |
| 1320 | 14.1 | 1 |

Transfer across the placenta of the cow has been demonstrated by Whiting et al. (1972). Dams were fed at dosages of 0.25 to 1.00 ppm in the feed. Calves analyzed at parturition were found to contain DDE at levels proportional to the dam's dietary intake. Rumsey et al. (1973) confirmed that the total DDT in stillborn calves was similar to the cows in fat, organs, and muscles. They also determined that the equilibrium of DDT concentrations in blood between the mother and calf was very rapid, occurring within 6 hours. $^{14}$C DDT was injected into pregnant heifers, and fetuses were taken by caesarean section. Residues in the fat and liver of the calf were also detected within 6 hours after DDT injection.

ACKNOWLEDGMENT

We thank Mr. Michael Ondik and Mr. Robert Mothersbaugh for handling and caring for the deer, feeding, sample taking, and observations. We thank Mrs. Kathy Malasics, Mrs. Susan Gage, and Mrs. Ida Harris for laboratory pesticide residue extractions. This work was supported by funds from USDI 14-16-0008-769. It was also supported by the Pennsylvania Agricultural Experiment Station (Journal Series No. 5184), authorized for publication October 20, 1976.

REFERENCES

Baetcke, K. P., Cain, J. D., Poe, W. E., *Pestic. Monit. J.* 6, 14 (1972).
Barrier, M. J., *Doctoral Dissertation,* Clemson Univ. (1970; *Diss. Abstr.* 71-9512, University Microfilms, Ann Arbor, Mich. (1970).
Benson, W. W., Smith, P., *Bull. Environ. Contam. Toxicol.* 8, 1 (1972).
Benson, W. W., Watson, M. , Wyllie, J., *Pestic. Monit. J.* 7, 97 (1973).
Benson, W. W., Gabica, J., Beecham, J., *Bull. Environ. Contam. Toxicol.* 11, 1 (1974).
Billing, K. J., Phelps, R. J., *Proc. Trans. Rhod. Sci. Assoc.* 55 (Pt. 1), 6 (1972).
Causey, K., McIntyre, S. C., Jr., Richburg, R. W., *J. Agric. Food Chem.* 20, 1205 (1972).
Cowan, R. L., Hartsock, E. W., Whelan, J. B., Long, T. A., Wetzel, R. S., *J. Wildl. Manage.* 33, 204 (1969).
Cranmer, M. F., Carroll, J. J., Copeland, M. F., *Bull. Environ. Contam. Toxicol.* 4, 214 (1969).
Davies, J. E., Edmundson, W. F., Maceo, A., Irwin, G. L., Cassady, J., Barquet, A., *Food Cosmet. Toxicol.* 9, 413 (1971).
Greenwood, R. J., Greichus, Y. A., Hugghins, E. J., *J. Wildl. Manage.* 31, 288 (1967).
Jewel, S. R., *Colorado Coop.Wildl. Res. Unit, Technical Paper. No. 8* (1967).
Leng, M. L., paper in this volume (1976).
Martin, W. L., Rogers, R. W., Essig, H. W., Pund, W. A., *J. Anim. Sci.* 42, 196 (1976).
Moore, G. L., Greichus, Y. A., Hugghins, E. J., *Bull. Environ. Contam. Toxicol.* 3, 269 (1968).
Morgan, D. P., Roan, C. C., *Arch. Environ. Health* 22, 301 (1971).
Pillmore, R. E., Finley, R. B., Jr., *Trans. North Am. Wildl. Nat. Resour. Conf.* 28, 409 (1963).

Reif, V. D., Littleton, B. C., Sinsheimer, J. E., *J. Agric. Food Chem.* 23, 996 (1975).
Reynolds, P. J., Lindahl, I. L., Cecil, H. C., Bitman, J., *J. Anim. Sci.* 41, 274 (1975).
Roan, C. C., Morgan, D. P., Paschal, E. H., *Arch. Environ. Health* 22, 309 (1971).
Rumsey, T. S., Samuelson, G., Bovard, K. P., Priode, B. M., *J. Anim. Sci.* 37, 1186 (1973).
Schwartz, C. C., Nagy, J. G., *J. Wildl. Manage.* 38, 531 (1974).
Sink, J. D., Varela-Alvarez, H., Hess, C., *J. Agric. Food Chem.* 20, 7 (1972).
Stanley, C. W., *J. Agric. Food Chem.* 14, 321 (1966).
Thompson, J. F., Ed., "Analysis of Pesticide Residues in Human and Environmental Studies," Perrine Primate Res. Labs., Environmental Protection Agency, Perrine, FL (1971).
Walker, K. C., George, D. A., Maitlen, J. C., *U.S. Dept. Agric. Res. Serv. Rep.* 33-105, 21 (1965).
Watson, M., Pharaoh, B., Wyllie, J., Benson, W. W., *Bull. Environ. Contam. Toxicol.* 13, 316 (1975).
Whiting, F. M., Brown, W. H., Stull, J. W., *J. Dairy Sci.* 55, 1499 (1972).
Whiting, F. M., Brown, W. H., Stull, J. W., *J. Dairy Sci.* 56, 1324 (1973).
Wilson, L. L., Kurtz, D. A., Ziegler, J. H., Rugh, M. C., Watkins, J. L., Long, T. A., Borger, M. L., Sink, J. D., *J. Anim. Sci.* 31, 112 (1970).
Wilson, L. L., Kurtz, D. A., Rugh, M. C., Chase, L. E., Ziegler, J. H., Varela-Alvarez, H., Borger, M. L., *J. Anim. Sci.* 33, 1356 (1971a).
Wilson, L. L., Kurtz, D. A., Rugh, M. C., Chase, L. E., Ziegler, J. H., Varela-Alvarez, H., Borger, M. L., *J. Anim. Sci.* 33, 1361 (1971b).

# THE FATE OF PHENYL $N,N'$-DIMETHYLPHOSPHORODIAMIDATE IN ANIMALS

R. L. Swann, M. W. Sauderhoff, D. A. Laskowski, and W. H. Braun

*Agricultural Products
and
Health and Environmental Research Departments
Dow Chemical U.S.A.*

ABSTRACT. *A pharmacokinetic and metabolism profile was determined for phenyl $N,N'$-dimethylphosphorodiamidate (proposed common name diamidafos) in Sprague Dawley (Spartan substrain) rats. The rats were administered 1 and 25 mg/kg of $^{14}C$-diamidafos by the intravenous route and 25 mg/kg $^{14}C$-diamidafos by the oral route. The $^{14}C$-activity was cleared rapidly from plasma by an apparent first order process ($t_{\frac{1}{2}}$ = 3 hours). The rate of clearance was independent of dose. The $^{14}C$-activity from orally administered material was rapidly absorbed and clearance from the plasma was similar to that observed for the intravenous dosage. Approximately 84% of the administered $^{14}C$-activity (oral and intravenous) was excreted into the urine during the first 36 hours. Approximately 4% was excreted over the next 36 hours. A pooled aliquot of the urine collected over the first 36 hour period was found to contain diamidafos and three metabolites, phenyl N-methyl-phosphorodiamidate, phenyl N-methyl-N-formylphosphorodiamidate, and conjugated phenol. Milk from lactating dairy cows fed diamidafos was also analyzed and found to contain these same metabolites, indicating both animals can metabolize diamidafos in a similar manner. The isolation of the intermediate metabolite phenyl N-methyl-N-formylphosphorodiamidate supports the metabolic sequence for secondary amines as proposed by Gaudette and Brodie (1959).*

INTRODUCTION

Control of nematodes in major crops by phosphorous containing compounds has been well demonstrated (Turner, 1963; Youngson and Goring, 1963). In 1959, Youngson of The Dow Chemical Company discovered a group of phosphorous compounds that were extremely effective in the control of root knot nematodes. The most active compound of this group was phenyl $N,N'$-dimethylphosphorodiamidate (proposed common name diamidafos).

$$\text{C}_6\text{H}_5\text{-O-P(=O)(NHCH}_3\text{)}_2$$

The physical properties of diamidafos are indicated in Table 1.

TABLE 1

*Physical properties of diamidafos*

| Solvent | Solubility (gm diamidafos/100 gm Solvent) |
|---|---|
| Acetone | 17.5 |
| Benzene | 1.5 |
| Carbon tetrachloride | 0.4 |
| Chloroform | 31.9 |
| Diethyl ether | 0.8 |
| Methanol | 177.0 |
| Methylene chloride | 17.6 |
| Water | 5.0 |
| Melting point | 105.5 - 106°C |
| Boiling point | 27°C/10 mm |
| Molecular weight | 200 |

Diamidafos is formulated as the active ingredient in Nellite[R] 90 and WS nematocide which is registered for use in tobacco transplant water for root knot nematode control. It is presently being field tested in combination with Lorsban[R] brand insecticides in the United States and Europe for the control of soil pests in tobacco, peanuts, soybeans and corn.

Extensive research has been conducted to determine the fate of diamidafos in selected plants, soil, and water (Meikle, 1968; 1976; Meikle and Austin, 1976; Meikle and Christie, 1969). The metabolic pathway is the same and is as follows:

$$\begin{array}{c}
\text{C}_6\text{H}_5\text{-O-P(=O)(NHCH}_3\text{)(OH)} + \text{CH}_3\text{NH}_2 \longrightarrow \text{C}_6\text{H}_5\text{-O-P(=O)(OH)}_2 + \text{CH}_3\text{NH}_2 \\
\uparrow \qquad \qquad \downarrow \\
\text{C}_6\text{H}_5\text{-O-P(=O)(NHCH}_3\text{)}_2 \qquad \text{HO-P(=O)(OH)}_2 + \text{CH}_3\text{NH}_2 \quad (+ \text{C}_6\text{H}_5\text{OH}) \\
\text{(diamidafos)} \qquad \uparrow \\
[\text{HO-P(=O)(NHCH}_3\text{)}_2] + \text{C}_6\text{H}_5\text{OH} \longrightarrow \text{HO-P(=O)(NHCH}_3\text{)(OH)} + \text{CH}_3\text{NH}_2 \\
(+ \text{C}_6\text{H}_5\text{OH})
\end{array}$$

In plants, no significant residues of diamidafos or any of its metabolites were found in the tobacco plant, cucumber vine, or cucumber vegetable at harvest time (Meikle, 1968; 1976). In soil studies it was concluded that diamidafos should be completely dissipated from the soil in approximately one growing season (Meikle and Christie, 1969).

These experiments prompted the examination of diamidafos

in animals, and studies were thus initiated to determine its fate in rats and in dairy cattle. This paper summarizes the results obtained to date.

RAT METABOLISM

*Experimental design.* The pharmacokinetic profile of diamidafos in laboratory animals helps in assessing its potential toxicity. A study was initiated using Sprague-Dawley (Spartan substrain) rats. $^{14}$C-Diamidafos was administered to rats at 1 and 25 mg/kg by the intravenous route and 25 mg/kg $^{14}$C-diamidafos by the oral route. The diamidafos was $^{14}$C-uniformly labeled with a specific activity of 2.15 µCi/mg. Plasma, urine, and feces were collected and analyzed for $^{14}$C activity.

*Pharmacokinetic results.* Radiochemical analyses of the plasma from rats dosed intravenously revealed that the $^{14}$C activity was cleared from plasma by an apparent first order process (Fig. 1). The rate constants for the clearance of $^{14}$C activity from plasma after the 1 and 25 mg/kg diamidafos intravenous dose were 0.225 hr$^{-1}$ ($t_{\frac{1}{2}}$, 3.08 hr) and 0.236 hr$^{-1}$ ($t_{\frac{1}{2}}$, 2.94 hr), respectively. There were no statistical differences between the rates of clearance at the two dose levels (t-test, p <0.05). Thus, clearance of $^{14}$C activity from plasma is not dose dependent within the range of the dose levels studied. In addition, statistical analysis (paired t-test) of the logarithmically transformed data revealed that the plasma concentration of $^{14}$C activity is not sex dependent.

The $^{14}$C-plasma/time profile revealed rapid absorption and clearance of radioactivity within 30 minutes after oral administration of 25 mg/kg $^{14}$C-diamidafos. The rate constants for clearance from the plasma are identical to the rate constants calculated for the same amount administered intravenously. The $^{14}$C plasma/time profiles are directly superimposable. Therefore, it is concluded that orally administered $^{14}$C activity was extensively if not completely absorbed. This point is reinforced by the fact that approximately 87% of $^{14}$C activity is found in the urine following administration of 25 mg/kg diamidafos by the oral route (Fig. 2).

The apparent volume of distribution for diamidafos, calculated by dividing the dose in mg/kg by the concentration of diamidafos at t=0, is not significantly different at the two dose levels. This indicates that the concentration of diamidafos in tissues will be proportional to the dosage.

Excretion of $^{14}$C activity in urine accounts for approximately 89% of the administered $^{14}$C activity after 1 and 25 mg/kg $^{14}$C-diamidafos. Approximately 7% of the $^{14}$C activity

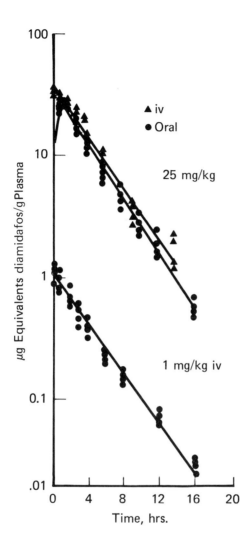

Fig. 1. Concentration of $^{14}C$ activity in plasma of rats versus time following administration of a 1 or 25 mg/kg, intravenous, or a 25 mg/kg oral dose of $^{14}C$-diamidafos.

is eliminated by biliary excretion and subsequent removal from the body in the feces. The excretion of $^{14}C$ activity from the body in urine is biphasic and essentially superimposable when percent of $^{14}C$ activity eliminated is plotted

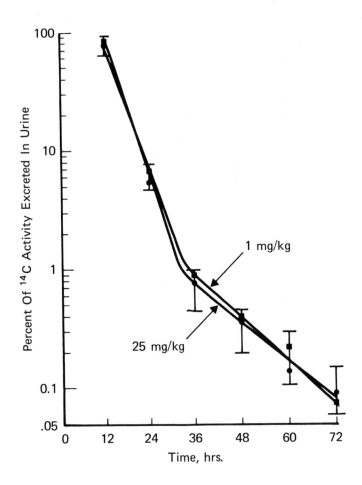

Fig. 2. Percent of total $^{14}C$ activity excreted in urine of rats after intravenous administration of 1 and 25 mg/kg $^{14}C$-diamidafos.

against time for the two dose levels. The rate constants for the excretion of $^{14}C$ activity at both dose levels via urine were not statistically different. Greater than 95% of the $^{14}C$ excreted was eliminated during the first 36 hours ($t_{1/2}$, 3-3.5 hr) after administration of the dose. Less than 2% of the administered dose was eliminated during the 36-72 hr time interval at both dose levels, and this later phase had a half-life of 10-11 hours (Fig. 2). Therefore, it does not seem likely that extensive accumulation would occur with repeated administration. There is no indication that significant amounts of $^{14}C$ activity (diamidafos and/or $^{14}C$ labeled metabolites) are retained in the body for long periods. Liver was the only tissue with detectable levels of $^{14}C$ activity at 72 hours after intravenous doses of 1 and 25 mg/kg $^{14}C$-diamidafos. The liver contained between 0.10 and 0.18 percent of the administered $^{14}C$ activity at this time. No $^{14}C$ activity was detected in the carcass (Table 2).

TABLE 2

Percent recovery of $^{14}C$ activity 72 hours after administration of $^{14}C$-diamidafos

| Dose, mg/kg | Route | Urine | Feces | Tissues | Carcass | Cage Wash | Total |
|---|---|---|---|---|---|---|---|
| 1 | iv | 89.63 ± 4.75 | 7.00 ± 1.23 | 0.10 ± 0.03 | <.05 | 0.94 ± 0.28 | 97.67 ± 5.42 |
| 25 | iv | 87.83 ± 2.78 | 7.01 ± 2.98 | 0.18 ± 0.04 | <.05 | 1.46 ± 0.80 | 96.47 ± 2.95 |
| 25 | Oral | 87.15 ± 1.94 | 7.96 ± 2.13 | 0.11 ± 0.04 | <.05 | 0.70 ± 0.21 | 95.92 ± 2.73 |

*Metabolite Identification.* Since greater than 95% of the $^{14}C$ urine excretion occurred within the first 36 hrs after the treatment, an aliquot of each urine sample from each animal (intravenously dosed) through the first 36 hours was pooled for analysis. The pooled urine was extracted with methylene chloride, concentrated and spotted onto 5 x 20 cm commercially precoated 0.25 mm thick silica gel TLC plates. The plates were developed in acetone, dried, and scanned on a Vanguard 880 plate scanner.

Four discrete radioactive areas were detected (Fig. 3). These same areas were found for urine samples of orally dosed animals. The radioactive areas on the TLC plates were scraped, collected into fritted glass columns, and then eluted with methanol. The methanol extracts were concentrated and chromatographed on a Tracor MT 222 gas chromatograph equipped with a flame photometric detector (phosphorous mode). The solutions corresponding to peaks I through III all gave positive phosphorous responses, while the solution corresponding to peak IV gave no response. The methanolic solutions were then chromatographed on a Finnigan Model 3100D gas chromatograph/mass spectrometer using both chemical ionization (CI) and electron impact (EI) systems. The mass spectra obtained for the compounds represented in TLC peaks I, II and III are shown in Figs. 4-6.

The molecular weights of the compounds in TLC peaks I, II and III as indicated by their chemical ionization mass spectra were 186, 200 and 214, respectively. The compound in

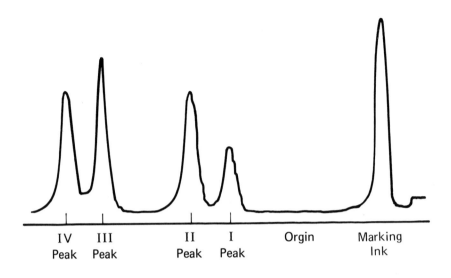

Fig. 3. TLC scan of methylene chloride extract of urine from rats dosed with $^{14}C$-diamidafos.

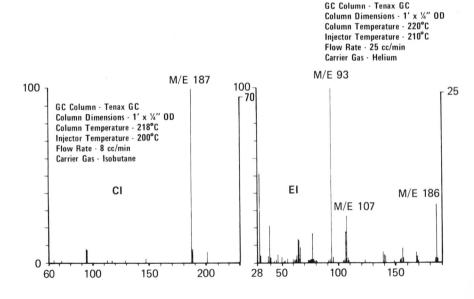

Fig. 4. Chemical ionization (CI) and electron impact (EI) mass spectra of TLC peak I.

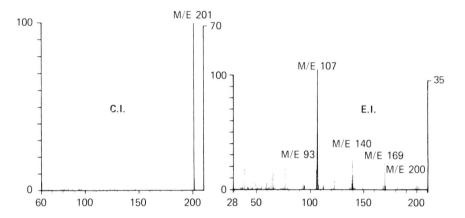

Fig. 5. Chemical ionization (CI) and electron impact (EI) mass spectra of TLC peak II.

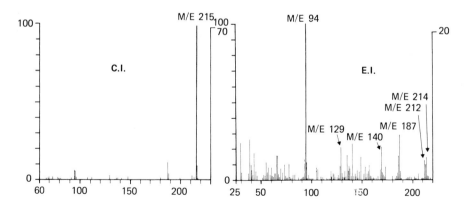

Fig. 6. Chemical ionization (CI) and electron impact (EI) mass spectra of TLC peak III.

TLC peak IV was not volatile enough to pass through the gas chromatograph, and a solid probe injection gave too complex a spectrum for interpretation. Its identity will be discussed later.

TLC peak II had the same TLC $R_f$, GLC retention time, and molecular weight as diamidafos. The EI and CI mass spectra obtained from the solution containing TLC peak II was essentially identical to the EI and CI mass spectra of diamidafos, and component II was concluded to be the unchanged parent compound.

Several structures might be predicted from the electron impact spectra of peaks I and III, shown in Figs. 4 and 6. To confirm the identity of the two materials, an IR spectrum and a high resolution mass spectrum were obtained from each. The samples used for IR determination were prepared by dilution of an aliquot of the methanolic solutions from the TLC plates with water. Then each was extracted several times with methylene chloride. The methylene chloride was concentrated, and then passed through a neutral alumina column. The compounds were eluted from the alumina column with water and repartitioned into methylene chloride. The methylene chloride extracts were evaporated to dryness and a KBr pellet made of the residue.

The IR spectrum obtained from TLC peak I is shown in Fig. 7, and a summary of the interpretation is given in Table 3.

Fig. 7. *IR spectrum of TLC peak I.*

Based on the IR, electron impact, and chemical ionization mass spectra (Fig. 4) data, the following structure for TLC peak I is proposed:

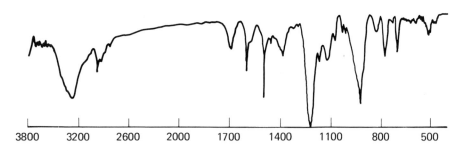

TABLE 3

*Summary of IR data from TLC peak I*

| Band Cm$^{-1}$ | Possible Group |
|---|---|
| 1598, 1497, 1076, 1020, 1011 | Phenyl Group |
| 775, 696 | Phenyl Group |
| 923 | O - P |
| 1220 | P = O |
| 2938, 2828 | N-CH$_3$ |
| 1120, 825 | P-N-CH$_3$ |
| 1400 | P-(NH) |
| 1574 | P-(NH$_2$) |

The IR spectrum obtained from TLC peak III is shown in Fig. 8, and a summary of the data is given in Table 4.

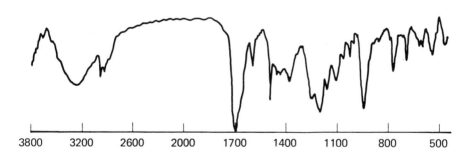

Fig. 8. IR spectrum of TLC peak III.

TABLE 4

*Summary of IR data obtained from TLC peak III*

| Band Cm$^{-1}$ | Possible Group |
|---|---|
| 1597, 1497, 1460, 1029, 1010 | Phenyl |
| 772, 694 | Phenyl |
| 1200 | P = O |
| 947 | O - P |
| 1167 | ∅ - O |
| 3250 | N - H |
| 1400 | $\overset{O}{\underset{\parallel}{P}}$-NH |
| 1702 | $\overset{O}{\underset{\parallel}{H-N-C-H}}$ |

With the IR data given above, along with the EI and CI mass spectra, the following structure for TLC peak III is proposed:

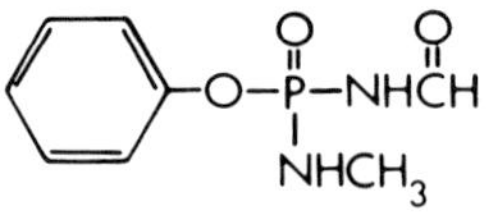

$$\text{C}_6\text{H}_5-\text{O}-\overset{O}{\underset{\underset{NHCH_3}{|}}{\overset{\parallel}{P}}}-\text{NH}\overset{O}{\overset{\parallel}{C}}\text{H}$$

A high resolution mass spectrum was obtained from each compound to determine if the proposed structures for the metabolites were consistent with their elemental composition. The high resolution mass spectra were obtained by chromatographing methanolic solutions containing the isolated metabolites on a AEI MS-30 high resolution GLC/mass spectroscopy system. The gas chromatographic parameters were identical to the ones given in Fig. 4. The results of these analyses were as follows: Metabolite I (theoretical MW = 186.0549; observed MW - 186.0566; difference = 9 ppm); Metabolite III (theoretical MW = 214.0504; observed MW = 214.0515; difference = 5 ppm). Thus, the IR, GLC, and mass spectral data support the proposed structures for the metabolites. The final confirmation step was achieved by synthesizing the two proposed compounds and comparing them with the metabolites.

The two compounds were synthesized by H. O. Senkbeil and R. B. Rogers of The Dow Chemical Company. EI and CI mass spectra were obtained for the two synthetic compounds and their spectra were compared to those obtained from the two metabolites.

In each case, the mass spectra showed excellent agreement between the metabolites and the synthesized compounds. The synthesized compounds were also compared to the metabolites by TLC and GLC, and in each case were found to have the same $R_f$ value on TLC and the same retention time by GLC. With these facts and the data presented earlier, it was concluded that the compounds synthesized were the same as the urinary metabolites isolated.

The compound in TLC peak IV, which was not volatile enough to pass through the gas chromatograph, was suspected to be a conjugated phenol. It was therefore refluxed in a 38% HCl solution for 4 hours, and at the end of the 4 hour hydrolysis the $^{14}C$ activity was extracted into ether. The product obtained from this hydrolysis was examined by TLC and determined to be phenol.

In summary, the four radioactive compounds found in the rat urine have been isolated and identified as follows:

phenyl-*N*-methylphosphorodiamidate (I),
phenyl *N,N'*-dimethyl phosphorodiamidate (II),
phenyl *N*-methyl-*N*-formyl phosphorodiamidate (III), and
conjugated phenol (IV).

CATTLE METABOLISM

A feeding study in which nonlabeled diamidafos was fed in the diet of lactating cows was performed to determine the amount of diamidafos and phenol present in the milk. The dosage of diamidafos was increased every two weeks until a maximum of 300 ppm was reached. Levels fed for the two-week intervals were 1, 3, 10, 30, 100 and 300 ppm. Milk from the lactating cows after 6 days at the 300 ppm feeding level was examined for the presence of the metabolites found in rat urine.

It was determined that metabolites I and III would go through the same clean-up procedure as the one used to analyze for diamidafos in milk (Swann, 1976). A large aliquot of the skim milk from cows fed at the 300 ppm feeding level for 6 days was prepared for analysis by this clean-up procedure. An aliquot of the final solution obtained from this method was injected into the Finnigan Model 3100D GLC/mass spectroscopy system, and EI and CI mass spectra were obtained from components in the solution. The analysis demonstrated that each of the phosphorus containing metabolites identified in rat urine were also present in the milk from lactating cows, suggesting that both animals metabolize diamidafos in a similar fashion.

DISCUSSION

It has been reported by Gaudette and Brodie (1959) that secondary alkylamines are metabolized in liver by the microsomes to form primary amines and formaldehyde. These authors also proposed that this type of N-dealkylation was the primary way in which animals detoxify secondary amines, but no direct proof was offered that the intermediates proposed were actually formed. The general equation used to express the reaction is the following:

$$R\text{-}NHCH_3 + [OH] \longrightarrow \underset{\text{Unstable Intermediate}}{[RNHCH_2OH]} \longrightarrow RNH_2 + H_2CO$$

With the metabolites found in the rat urine, and milk from lactating dairy cows, the following general metabolic pathway is proposed for diamidafos:

It is noted that the isolation of the intermediate metabolite phenyl N-methyl-N-formylphosphorodiamidate supports the metabolic sequence for secondary amines in animals as proposed earlier (Gaudette and Brodie, 1959).

SUMMARY

Diamidafos is rapidly absorbed into the blood stream when animals are given an oral dose. Once in the blood stream, either by the oral or the intravenous route, diamidafos is rapidly metabolized, most likely in the liver. The metabolites, along with some unmetabolized diamidafos, are filtered out in the kidneys and excreted into the urine.

Milk from lactating cows fed diamidafos was analyzed and found to contain the same phosphorus containing metabolites that were present in the rat urine, indicating that both animals metabolize diamidafos in a similar manner.

It is proposed that diamidafos is metabolized by the same pathways as suggested by Gaudette and Brodie (1959) for secondary alkylamine compounds.

ACKNOWLEDGEMENT

We thank the personnel in the Agricultural Products Department, Dow Chemical USA, and Carol Swann for support in the preparation of this paper.

REFERENCES

Gaudette, L. E., Brodie, B. B., *Biochem. Pharmacol.* 2, 89 (1959).
Meikle, R. W., *J. Agric. Food Chem.* 16, 928 (1968).
Meikle, R. W., The Dow Chemical Company, personal communication (1976).
Meikle, R. W., Austin, S. J., The Dow Chemical Company, personal communication (1976).
Meikle, R. W., Christie, P. H., *Bull. Environ. Contam. Toxicol.* 4, 88 (1969).
Nyquist, R. A., The Dow Chemical Company, personal communication (1976).
Sauerhoff, R. H., The Dow Chemical Company, personal communication (1976).
Swann, R. L., unpublished data (1976).
Turner, G. O., *Down to Earth* 18, 16 (1963).
Youngson, C. R., Goring, C. A. I., *Down to Earth* 18, 3 (1963).

# METABOLISM OF CRONETON
## (2-ETHYLTHIOMETHYLPHENYL *N*-METHYLCARBAMATE)
## IN LARGE ANIMALS

H. W. Dorough and D. E. Nye[1]

Department of Entomology
University of Kentucky

*ABSTRACT. A lactating Holstein cow and a male Yorkshire pig treated with a single oral dose, 0.5 mg/kg, of ring [$^{14}$C] Croneton excreted 97.8% and 90.0% of the dose via the urine after 24 hr. Residues were not detected in swine tissues sampled 24 hr after treatment, and of the bovine tissues, only the kidney, liver and skin contained detectable radiocarbon (0.016, 0.017, and 0.05 ppm $^{14}$C-Croneton equivalents, respectively). Milk collected 6 hr after treatment contained 128 ppb [$^{14}$C] residues; 60% was as the free carbamate metabolites, Croneton sulfoxide and sulfone. White Leghorn hens given ring [$^{14}$C] Croneton as a single oral dose (0.5 mg/kg) or as daily treatments for 7 days (0.5 mg/kg twice daily at 12-hr intervals) exhibited patterns of excretion and metabolism similar to those observed in the cow and pig. Birds sacrificed 4 hr after the last of the daily doses contained [$^{14}$C] Croneton equivalents ranging from 0.019 ppm in the fat to 0.324 ppm in the kidney. By 24 hr, only the liver and kidney (0.044 and 0.022 ppm) contained residues in excess of 0.01 ppm, and these had declined to 0.01 ppm or less by 4 days. Residues in eggs were on the order of 0.03-0.04 ppm [$^{14}$C] Croneton equivalents after 2 days of treatment and reached a maximum of 0.06-0.07 ppm after 7 days. They declined rapidly when treatment was terminated and were below detectable levels after 3 days. About 75% of the radiocarbon in the eggs was as free phenol sulfoxide and sulfone, 10% as free Croneton sulfoxide and Croneton sulfone, 5% as water-soluble metabolites and 5% as unextracted residues. The remaining 5% was unknown metabolites in the organoextractable fraction.*

---

[1]Present affiliation: Thompson-Hayward Chemical Company.

INTRODUCTION

Croneton (2-ethylthiomethylphenyl-$N$-methylcarbamate) is an experimental insecticide which may be used on crops constituting a portion of the diets of livestock. Therefore, it is important to determine the fate of the insecticide in these animals so that the potential for the transfer of residues to the diet of man can be estimated. In the present study, results of the fate of a single oral dose of radioactive Croneton to a lactating Holstein cow, a male Yorkshire pig, and to laying White Leghorn hens are presented. Experiments with the hens were extended to include daily treatments with Croneton for 7 days so that the effect of multiple exposure on metabolism and disposition of the carbamate could be evaluated. The fate of Croneton in rats following single and multiple treatments of the insecticide has been determined in our laboratory and the results previously reported (Nye et al., 1976).

MATERIALS AND METHODS

*Chemicals*

To determine the fate of Croneton in the animals, [$^{14}$C]-ring-labeled insecticide was used (specific activity, 2.0 mCi/mmol). This was supplied by the Chemagro Corporation and had a radiochemical purity of greater than 99% as determined by thin-layer chromatographic and autoradiographic techniques. To aid in metabolite identification by cochromatography, the following compounds were synthesized by methods previously described (Nye et al., 1976); 2-ethylsulfinylmethylphenyl $N$-methylcarbamate (Croneton sulfoxide), 2-ethylsulfonylmethylphenyl $N$-methylcarbamate (Croneton sulfone), 2-ethylthiomethylphenol (Croneton phenol), 2-ethylsulfinylmethylphenol (phenol sulfoxide), and 2-ethylsulfonylmethylphenol (phenol sulfone). Structures of these compounds are shown in Fig. 1.

*Treatment and Sampling*

*Cow and pig.* Both the lactating Holstein cow (weighing approximately 500 kg with milk production of 50 lb/day) and the male Yorkshire pig (approximate weight of 50 kg) were acclimated to the experimental ambience for 3 days prior to treatment. Croneton was administered at a level of 0.5 mg/kg in a gelatin capsule containing feed grain. Following treatment, the cow was placed in a metabolism stall and separation

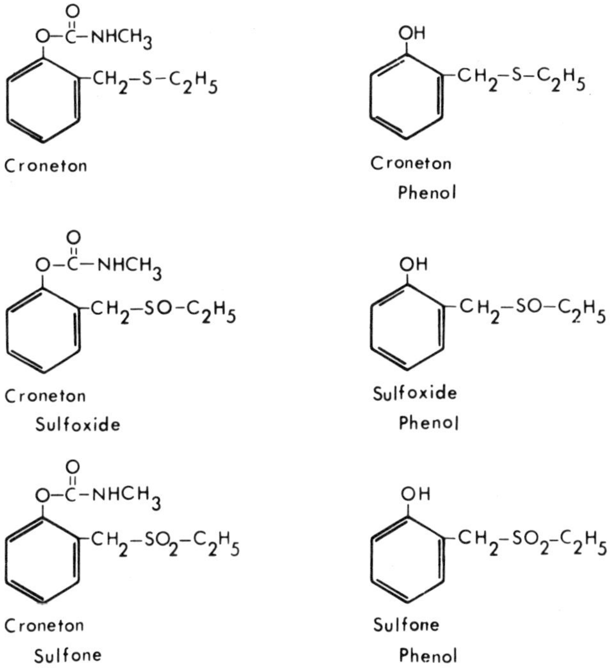

## Fig. 1. Structures of Croneton and metabolites.

Structures shown: Croneton, Croneton Phenol, Croneton Sulfoxide, Sulfoxide Phenol, Croneton Sulfone, Sulfone Phenol.

of urine from feces was accomplished by an intra-urethral catheter implanted prior to treatment. The swine was also placed in a metabolism stall which allowed for separate collection of urine and feces.

Urine eliminated by the cow was sampled on an hourly basis for the first 6 hr after treatment, at 2 hr intervals for the next 6 hr, and at 4 hr intervals for the next 12 hr. Urine eliminated by the pig was sampled periodically during the 24 hr experiment and collection was dependent on the intervals of urination. Radioassay of the urine samples was by direct liquid scintillation counting of 0.5 ml samples. Feces samples were collected from both animals periodically and radioassayed by combustion of 1.0 g samples using a Packard Model 306 Sample Oxidizer.

In addition to the urine and feces samples, blood (5.0 ml) and milk from each quarter (5-10 ml) were taken from the cow at the same time intervals as the urine samples. At 6 and 21 hr after treatment, the animal was milked in a normal manner using a mechanical milker. The blood samples, taken from the abdominal vein, were immediately heparinized and later radioassayed by combustion of 0.5 ml samples as described above.

At 24 hr post-treatment, the cow and pig were slaughtered

and representative tissue samples immediately frozen for later analysis. Radioassay of 0.5 g samples (0.05 g in the case of adipose tissue) was accomplished by combustion in the same manner as for the blood and feces samples.

Hens. In the first study, eight hens were treated with a single oral dose of Croneton at a rate of 0.5 mg/kg (sp. act., 2.0 mCi/mmol). The dose was prepared by adding the ring labeled insecticide to gelatin capsules containing a small amount of laying mash. Following treatment, the birds were placed in individual cages and eggs and excreta were collected at 12-hr intervals for 3 days. No attempt was made to separate urine and feces. At 8 hr, and at 1, 2, and 3-day intervals after treatment, two birds were sacrificed and tissue samples collected.

In the second study, 10 birds were given the capsules containing Croneton on a twice-daily basis at about 12-hr intervals for 7 days. The specific activity of the ring [$^{14}$C] material was 4.0 mCi/mmol to facilitate tissue radiocarbon analysis which was difficult at the dosage level used in the first study. Eggs and excreta were collected prior to treatment and on a daily basis after treatment was initiated. On the 4th day of treatment and prior to administration of the second daily dose, one bird was sacrificed and tissues taken for radioassay. At 4 hr following the final treatment, four birds were sacrificed in an attempt to achieve maximum tissue [$^{14}$C]-ring Croneton equivalents. Two birds were sacrificed 1 day after the final treatment and one bird sacrificed at 2, 4, and 7 days after the last capsule was given to establish rates of dissipation of [$^{14}$C]-ring Croneton residues.

In all experiments, the birds were maintained on a 12-hr light and 12-hr dark photo period and were given laying mash and water *ad libitum*. All samples were stored in the freezer until analyzed.

Tissue samples and the excreta were radioassayed by combustion as previously described. The eggs also were analyzed by combustion, but they were first divided into shell, yolk and whites, and analyses were performed using 0.2-g samples.

## Metabolite Identification

*Excreta.* Aliquots of urine from the cow and swine, or excreta from the hens were lyophilized prior to the determination of the nature of the radiocarbon content. The dry solid residues were then repeatedly washed with methanol which removed greater than 95% of the [$^{14}$C]-ring Croneton equivalents while leaving a large portion of the interfering components behind. The methanolic fraction was concentrated to a volume suitable for direct application to silica gel thin layer chro-

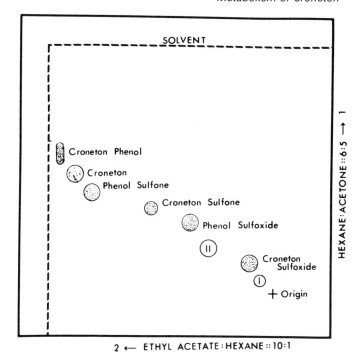

*Fig. 2. Drawing of TLC showing relative positions of Croneton and its metabolites.*

matographic plates. Two basic solvent systems were used to develop the TLC plates; a 6:5 hexane-acetone mixture for 1-dimensional chromatography, and a 2-dimensional system using 6:5 hexane-acetone in one direction and 10:1 ethyl acetate-hexane in the second direction (Fig. 2). The first system was used as a preparative method for analysis of free metabolites, while the 2-dimensional system was used for attempted co-chromatography of metabolites with known standards. In addition, a benzene-acetonitrile solvent system (of varying composition) was used to confirm the results of the 2-dimensional system.

The polar metabolites remaining at the origin in the above systems were subjected to analysis by incubation with 0.5 N hydrochloric acid. The acidified samples were incubated for 6 days at 45°C and the acid-released metabolites were removed by extraction with chloroform at 24-hr intervals.

In addition to acid hydrolysis, the polar metabolites were incubated with β-glucuronidase (type V-A, Sigma Chemical Company) and aryl sulfatase (type III, Sigma Chemical Company). Fifty units of β-glucuronidase or 10 units of aryl sulfatase were added to the extracted metabolites in acetic acid-sodium acetate buffer (0.1 M, pH 5.6) and the mixture incubated for

24 hr at 37°C. The preparations were then extracted with three equal volumes of chloroform, and the chloroform concentrated and applied to TLC.

*Milk.* To determine the nature of the radiocarbon residues in milk, 100 ml of the 6-hr sample were mixed with 200 ml of acetone to precipitate the solids. The solids were separated by filtration and the filtrate extracted three times (100 ml, 75 ml, and 75 ml) with chloroform. The combined chloroform extracts were concentrated and analyzed by TLC using the 6:5 hexane-acetone solvent system.

The aqueous phase of the milk was acidified to 1.0 N with hydrochloric acid and heated for 1 hr at 80°C. Radiocarbon released by acid treatment was extracted into chloroform and analyzed by TLC.

*Eggs.* Eggs collected during the continuous feeding study were analyzed by homogenizing the combined whites and yolks in acetonitrile for 3 min in a Lourdes homogenizer. The solids were removed by filtration and twice again homogenized with acetonitrile. The acetonitrile filtrates were combined and 100 ml of 0.05 N hydrochloric acid were added. This mixture was extracted with four 60-ml aliquots of chloroform, the chloroform dried over anhydrous sodium sulfate, and then evaporated to dryness. Several washings of the evaporating flask with a precipitation solution (1.25 g of ammonium chloride and 25 ml of phosphoric acid in sufficient water to make a final volume of one liter) were used to transfer the residue to a separatory funnel. The solution was extracted as before with chloroform, the solvent dried with sodium sulfate and then concentrated for application to TLC plates.

The water resulting from the first chloroform extract was acidified to 1.0 N with hydrochloric acid and heated at 80°C for 1 hr. This solution was extracted with chloroform. Solids remaining after homogenization were similarly treated with a 1.0 N hydrochloric acid solution for 1 hr at 80°C and the solution filtered. The filtrate was extracted with chloroform and both phases were radioassayed.

RESULTS

*Cow and Pig.*

*Uptake and excretion.* Maximum concentrations of [$^{14}$C]-ring Croneton equivalents in cow blood, 0.32 ppm, occurred 3 hr after treatment (Fig. 3). The appearance of radiocarbon residues in the milk was somewhat slower with the maximum concentration of 0.15 ppm observed 4 hr after administration of the capsule. These results are similar to those observed in rats

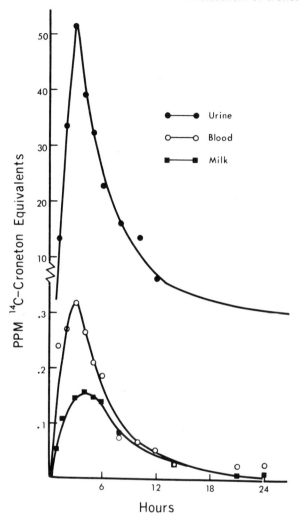

Fig. 3. Ppm [$^{14}C$] ring-Croneton equivalents in urine, blood, and milk of a lactating Holstein cow treated with a single oral dose, 0.5 mg/kg, of the insecticide.

where blood and tissue [$^{14}C$]-Croneton equivalents were highest after 2.0 to 2.5 hr (Nye et al., 1975). [$^{14}C$]-Croneton equivalents in the urine reached a maximum concentration 3 hr after administration. Cumulative excretion of [$^{14}C$] Croneton equivalents via the urine was greater than 90% of the dose by 12 hr (Fig. 4). After 24 hr, 97.8% of the dose had been voided in the urine while less than 1% was in the feces. In the pig, elimination was also principally via the urine, with 80% of

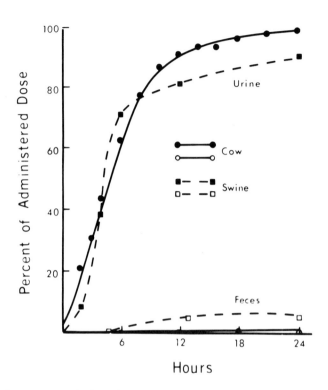

*Fig. 4. Elimination of radiocarbon from a cow and pig treated with a single oral dose, 0.5 mg/kg, of [$^{14}$C] ring-Croneton.*

the dose excreted within the first 12 hr. After 24 hr, 90% of the dose had been eliminated in the urine and 5.1% was present in the feces.

*Tissues.* Following slaughter of the swine 24 hr after treatment, combustion analysis of various tissues revealed no detectable radioactive residues. Essentially the same findings were obtained with the bovine tissues. Of 18 different tissues analyzed, only the kidney (0.016 ppm), liver (0.017 ppm), and skin (0.05 ppm) contained detectable levels of radiocarbon.

*Urinary metabolites.* There were several quantitative differences in the nature of the Croneton metabolites in the cow and pig urine (Table 1). The cow, for example, excreted Croneton principally as water soluble metabolites, 86.9% of the dose, with the phenol sulfoxide and phenol sulfone being the only significant free metabolites. The swine, on the other hand, eliminated the [$^{14}$C]-ring-Croneton equivalents to a lesser extent as water soluble metabolites (53.5% of the

TABLE 1

Nature of radiocarbon in 0-12 hour urine of cow and pig treated with a single oral dose, 0.5 mg/kg, of $[^{14}C]$ ring-Croneton

| Metabolites[1] | | % of total $[^{14}C]$ in sample | |
|---|---|---|---|
| F - Free<br>A - Acid-released<br>G - Glucuronidase-released<br>S - Sulfatase-released | | Cow | Pig |
| Croneton | - A | 0.0 | 9.0 |
| Croneton Sulfoxide | - F | .5 | 0.0 |
|  | - A | .9 | 0.0 |
|  | - G | 0.0 | 2.1 |
|  | - S | 0.0 | 2.0 |
| Croneton Phenol | - F | 0.0 | 25.5 |
| Phenol Sulfoxide | - F | 8.7 | 9.1 |
|  | - A | 12.5 | 12.2 |
|  | - G | 37.9 | 12.1 |
|  | - S | 4.8 | 18.0 |
| Phenol Sulfone | - F | 2.4 | 6.0 |
|  | - A | 1.0 | 1.0 |
|  | - G | 6.3 | 4.3 |
|  | - S | 1.8 | 4.3 |
| Unknown II | - A | 7.5 | 0.0 |
| Polar Unknown[2] | - A | 66.5 | 33.6 |
|  | - G | 44.2 | 40.9 |
|  | - S | 81.8 | 35.1 |

[1] After removing the free metabolites, the polar radiocarbon was separated into 3 portions and individually treated with acid, glucuronidase or sulfatase. Therefore, the released metabolites are not cumulative.

[2] Radiocarbon remaining in water phase after indicated treatment.

dose). Of the total radiocarbon in the pig urine, 26% was as free Croneton phenol, 9% as free phenol sulfoxide and 6% as free phenol sulfone.

Analysis of the polar urinary [$^{14}$C]-components, those remaining at the TLC origin, was attempted by acid hydrolysis. In the case of urine collected from the cow, incubation at 45°C for 6 days in 0.5 N HCl or at 80°C for 2 hr at 4.0 N HCl, gave identical results. Only 25% of the polar metabolites was recovered by chloroform extraction. Nearly one-half of the radiocarbon released with acid was identified as the phenol sulfoxide. The remainder consisted of Croneton sulfoxide, phenol sulfone, Croneton, phenol, and unknown II. The latter metabolite had the same characteristics as a metabolite formed in rats (unknown A, Nye et al., 1976) and in chickens (unknown II). Incubation of the polar metabolites with β-glucuronidase was more successful at releasing the [$^{14}$C]-Croneton equivalents from the polar form than treatment with acid (Table 1). TLC analysis revealed that conjugated phenol sulfoxide constituted 37.9% of the radiocarbon in the urine and the phenol sulfone, 6.3%. Aryl sulfatase treatment released only 4.7% and 1.7% as the phenol sulfoxide and phenol sulfone, respectively. In neither case did the enzyme treatment release unknown II.

Only 33.6% of the Croneton equivalents in the swine urine remained as water soluble components after treatment with acid (Table 1). About 9% of the total [$^{14}$C] in the urine was released as phenol sulfoxide, while only 1% was released as the phenol sulfone. In addition, 9% was released as a product with an $R_f$ identical to Croneton. Unknown II was not observed in the acid released fraction of the swine urine.

Enzymatic hydrolysis of the polar origin Croneton equivalents was only slightly more effective than treatment with acid, and the radiocarbon released was nearly equal for glucuronidase and sulfatase. As was the case for the cow urine, the major component released by enzymatic hydrolysis was the phenol sulfoxide. After glucuronidase treatment, 41% of the radioactivity in the urine remained in the water phase while 35% remained after treatment with sulfatase.

It appears that the differences that occurred between the two animals with regard to the fate of Croneton was their ability to oxidize the parent carbamate as well as to conjugate the phenolic metabolites of Croneton. The cow had the greater ability to perform these functions, while the swine had a somewhat restrictive ability to form conjugates of the products resulting from hydrolysis of the carbamate. It is important, however, that this did not limit the swine's ability to eliminate the ingested compound.

*Milk metabolites*. While no appreciable radiocarbon was detectable in the milk 21 hr following administration of the bolus, analysis of the metabolites in the 6-hr sample (0.3% of the dose) gave a good indication of the nature of the products excreted in milk. Of the total radiocarbon in the milk, 67.3% was organosoluble and identified primarily as the sulfoxide and sulfone of the carbamate (Table 2). Small amounts of the phenol sulfoxide and sulfone were also present in the free form.

Acid hydrolysis of the water soluble fraction yielded organo extractables primarily as Croneton sulfoxide and phenol sulfoxide. Other minor products identified were Croneton phenol, phenol sulfone, and Croneton sulfone. An unknown metabolite (II), tentatively identified as $N$-hydroxymethyl Croneton

TABLE 2

Nature of $[^{14}C]$ Croneton equivalents in milk of cow treated with a single oral dose, 0.05 mg/kg, of $[^{14}C]$ ring-Croneton

| Metabolites or fraction | PPB and % distribution/6-hr milk sample | |
|---|---|---|
| | PPB | % |
| Free | 86.1 | 67.3 |
| Croneton Sulfoxide | 50.6 | 39.6 |
| Croneton Sulfone | 24.7 | 19.3 |
| Phenol Sulfoxide | 7.2 | 5.6 |
| Phenol Sulfone | 3.6 | 2.8 |
| Water Soluble | 37.8 | 29.6 |
| Acid-released | 15.7 | 12.3 |
| Croneton Sulfoxide | 7.2 | 5.6 |
| Croneton Sulfone | .4 | .3 |
| Croneton Phenol | .6 | .5 |
| Phenol Sulfoxide | 4.8 | 3.8 |
| Phenol Sulfone | 1.5 | 1.2 |
| Unknown II | 1.2 | .9 |
| Remaining in water | 22.1 | 17.3 |
| Milk Solids | 4.0 | 3.1 |
| Total | 127.9 | 100.0 |

sulfone, was detected in milk analyzed shortly after collecting the sample, but was not observed in samples stored after 9 months. Other metabolites were stable in the frozen milk. Initially, unknown II represented 0.9% of the total radiocarbon in the milk or 1.2 ppb [$^{14}$C] Croneton equivalents. Water soluble materials which were not acid labile constituted 17.3% of the [$^{14}$C] residues in the milk, while only 3.1% was associated with the solids precipitated by the addition of acetone.

*Hens*

*Excretion and distribution of residues.* Elimination of the single oral dose of Croneton to White Leghorn laying hens was very rapid with greater than 80% of the dose voided in the excreta within 24 hr, and greater than 90% after 3 days. A somewhat different pattern was observed in the 7-day study (Fig. 5). When the insecticide was administered on a continu-

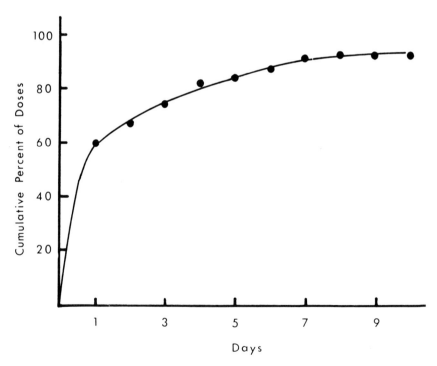

*Fig. 5. Elimination of radiocarbon in the excreta of hens treated orally with [$^{14}$C] ring-Croneton every 12 hr, 0.5 mg/kg, for 7 days.*

ous basis, elimination via the excreta was somewhat low for the first 2 days following initiation of treatment with 60% of the cumulative dose being excreted; however, 92.0% of the radiocarbon consumed over the 7-day period had been eliminated when treatment was terminated.

Results of the radioassay of selected tissue samples collected after administration of both a single oral dose, or continuous treatments of the insecticide are given in Table 3. At 8 hr after the single dose, slight tissue residues were present but dissipation of Croneton equivalents from the tissue was very rapid and no detectable [$^{14}$C]-equivalents were found 24 hr post-administration.

Accumulation of insecticide equivalents in the tissues was minimal as indicated by the low levels found in isolated tissues 4 days after initiation of the daily treatments. Although [$^{14}$C]-residues were higher in the tissues at 7 days, this was due largely to the early sampling time after the last dose was administered (4 hr) rather than from accumulation. With the 4-day samples, the hens were sacrificed 8 hr after the prior treatment. [$^{14}$C]-Residues in all tissues dissipated very rapidly and most were below the level of sensitivity (.005 ppm) within 2 days after treatment was terminated.

Eggs collected following the single oral treatment contained no [$^{14}$C]-ring Croneton equivalents in either shell, white, or yolk. [$^{14}$C]-Residues in eggs collected during the continuous feeding study were evident 2 days after initiation of treatment. The shell showed the lowest level of insecticide residues, reaching a high of 16.1 ppb on the third day. Residues in the white and yolk were nearly equal for the first 4 days of treatments and averaged 38 ppb. While residues in the white increased very little thereafter, those in the yolk continued to rise and were on the order of 70 ppb on the last day of treatment.

A large variation existed between the eggs sampled at any given time during the continuous feeding experiment. The location of residues within the eggs appeared to be dependent on the stage of egg development at the time the dose was administered and certain eggs within each sampling time were found to contain the entire residue within either the white or the yolk.

*Nature of metabolites.* The nature of the [$^{14}$C]-ring-Croneton equivalents found in the excreta is given in Table 4. The differences in metabolism that existed between application by single oral dose or continuous feeding occurred in two areas. First, the quantity of polar materials in the excreta of the single-dosed hens was lower than that observed in the continuous feeding experiment. This difference was minimized after acid hydrolysis, as similar amounts of [$^{14}$C]-ring Croneton equivalents remained as unknown polar products in

TABLE 3

[$^{14}C$] Ring Croneton equivalents in tissues of hens administered the insecticide as a single oral dose or as twice daily treatments for 7 days (0.5 mg/kg per dose).

PPM Croneton Equivalents

| Tissue | Single oral dose[1] 8 hr | On treatment[2] | | Daily treatment | | Off treatment | |
|---|---|---|---|---|---|---|---|
| | | 4 days | 7 days | 1 day | 2 days | 4 days | 7 days |
| Blood | 0.053 | 0.016 | 0.099 | 0.010 | 0.009 | ND | ND |
| Brain | .028 | .001 | .039 | ND | ND | ND | ND |
| Fat | ND | ND | .019 | ND | ND | ND | ND |
| Heart | ND | .005 | .072 | .004 | .005 | ND | ND |
| Kidney | .062 | .016 | .324 | .022 | .018 | .007 | .006 |
| Liver | .035 | .025 | .180 | .044 | .034 | .009 | .011 |
| Muscle | .039 | ND | .069 | .004 | .006 | .001 | .002 |
| Skin | .022 | .013 | .075 | .007 | .012 | .006 | .004 |

[1] Tissues were analyzed at 1, 2, and 3 days after treatment, but did not contain detectable levels of residues.

[2] Higher residues at 7 vs. 4 days was due largely to the fact that the 7-day birds were sacrificed 3-4 hr after last treatment while 4-day birds were sacrificed 8 hr after last treatment.

TABLE 4

Nature of radiocarbon in excreta of hens treated with $[^{14}C]$-ring labeled Croneton.

| Metabolites | | % of total $[^{14}C]$ in sample | | | |
|---|---|---|---|---|---|
| F = Free | | Single oral dose | Continuous treatments/Days | | |
| A = Acid-released | | | 1 | 4 | 7 |
| Croneton Sulfoxide | - F | 0 | 2.7 | 1.3 | 1.7 |
| | - A | 0 | 0 | 1.2 | 0 |
| Croneton Sulfone | - F | 0 | .9 | .9 | .7 |
| Croneton Phenol | - F | 2.9 | 1.3 | 2.6 | 2.5 |
| Phenol Sulfoxide | - F | 23.9 | 2.6 | 5.1 | 6.0 |
| | - A | 1.0 | 9.8 | 2.6 | 3.4 |
| Phenol Sulfone | - F | 36.5 | 13.3 | 29.3 | 20.2 |
| | - A | 1.6 | 15.1 | 12.3 | 14.1 |
| Unknowns | - F | 0 | 12.1 | 9.6 | 10.6 |
| | - A | 7.1 | 7.2 | 9.4 | 13.3 |
| Unknown Polar[1] | | 27.0 | 35.0 | 25.7 | 27.4 |

[1] Radiocarbon remaining in water phase after acid treatment.

both studies. A second difference was in the nature of the radiocarbon which existed as free metabolites, and which could be released with acid. Like the single oral dose experiment, the major metabolites in the excreta of the multiple-dosed hens were the phenol sulfoxide and phenol sulfone. However, conjugation of these products was much greater in the continuous feeding study. Some small amounts of the sulfoxide and sulfone of the carbamate were also found in the continuous feeding study, but not in the single dose experiment.

Major differences in the quantity of unknown metabolites were found in the two studies (Table 4). The relative mobility of these unknowns on silica gel TLC plates are shown in Fig. 2. Unknown II (10.6% of the dose) has the same $R_f$ as an unknown isolated and tentatively identified as the $N$-hydroxymethyl Croneton sulfone in previous rat studies (Nye et al.,

1976). As found in the earlier study, unknown II in the chicken excreta was unstable, and was converted to the phenol sulfone under acid conditions. Several other unknowns were also released by acid hydrolysis but were not identified because of their low individual quantity. The major among these was unknown I and chromatographed as shown in Fig. 2. Treatment of the water soluble metabolites with β-glucuronidase and aryl sulfatase did not release any [$^{14}$C]-ring Croneton equivalents into the chloroform extract.

The nature of [$^{14}$C] residues in the tissues was not evaluated. Total radiocarbon was either too low in tissues of bulk, or the sample too small to yield sufficient [$^{14}$C] to follow through the various fractions resulting from extraction and partitioning. This was not the case with eggs (Table 5).

TABLE 5

Nature of [$^{14}$C] Croneton equivalents in eggs of hens dosed every 12 hr with 0.5 mg/kg [$^{14}$C] ring Croneton for 7 days

| Metabolites or fractions | PPB and % distribution in eggs laid on day | | | |
|---|---|---|---|---|
| | 3 | 6 | 7 | |
| | PPB | PPB | PPB | % |
| Free | 41.82 | 48.99 | 63.34 | 88.5 |
| Croneton Sulfoxide | 2.01 | 2.78 | 2.77 | 3.9 |
| Croneton Sulfone | 3.68 | 3.14 | 2.89 | 4.0 |
| Phenol Sulfoxide | 9.79 | 11.05 | 14.04 | 19.6 |
| Phenol Sulfone | 24.00 | 30.89 | 42.04 | 58.7 |
| Unknowns | 2.34 | 1.13 | 2.60 | 2.3 |
| Water Soluble | 2.25 | 2.70 | 3.35 | 4.7 |
| Acid-released | .57 | .69 | .85 | 1.2 |
| Remaining in water | 1.68 | 2.01 | 2.50 | 3.5 |
| Unextractables | 3.27 | 3.92 | 4.86 | 6.8 |
| Acid-released | .50 | .63 | .79 | 1.1 |
| Remaining in solids | 2.77 | 3.29 | 4.07 | 5.7 |
| Total | 47.34 | 55.61 | 71.55 | 100.0 |

A majority of the [$^{14}$C] residues in the eggs was organosoluble and, as was the case for metabolites in the excreta, those in the eggs were principally sulfoxidation products of Croneton phenol. Combined, the sulfoxide and sulfone of the phenol accounted for 71.4, 75.5, and 78.3% of the total residues in the eggs collected on day 3, 6, and 7. Small amounts of the oxidized carbamates and unknown products were also detected. [$^{14}$C] Residues in the water phase and solid fraction contained a combined total of 5 to 10% of the total radiocarbon in the eggs.

DISCUSSION

The carbamate insecticides are generally characterized as being efficiently metabolized and rapidly excreted by animals (Kuhr and Dorough, 1976). This very desirable feature has been of paramount importance in keeping alive the search for carbamates of commercial value during the past ten years. While progress was slow in the 1960's, several carbamates have more recently achieved commercial success in the Unites States and in other countries, and this success has stimulated greater research efforts on this group of insecticides.

Croneton, a Bayer product designated by the company initially as BAY HOX 1901, is one of the newer carbamate insecticides and one whose metabolism in animals has been studied extensively in our laboratory. Results obtained in the present study with the cow, pig, and hen are similar to those found with the rat (Nye et al., 1976). However, species differences do exist which are both of interest and of possible practical significance. Differences, as well as similarities, can best be demonstrated by considering the radiocarbon in the excreta of animals treated with a single oral dose of [$^{14}$C] ring Croneton (Table 6).

It is evident that the efficiency of excretion of [$^{14}$C] Croneton equivalents over a 24-hr period following treatment varied little among the species tested. In every case, 90% or more of the dose had been excreted. A major difference in the nature of the excreted radiocarbon was that the rat urine contained rather large quantities, 24% of the dose, of free carbamates, mostly Croneton sulfoxide. None of the large animals eliminated any free metabolite with the carbamate ester linkage intact. The pig and hen, however, did excrete much of the dose as free phenolic analogs of Croneton. They were minor constituents in cow and rat urine. Excretion of the free metabolites by the pig and hen, and the corresponding low levels of water soluble products demonstrated that conjugation is not always necessary for efficient elimination of phenolic metabolites of the carbamates.

TABLE 6

*Comparative nature of residues in the urine of animals treated with a single oral dose of [$^{14}$C]-ring labeled Croneton*

| Metabolites | % of dose as indicated metabolite | | | |
|---|---|---|---|---|
| | Rat | Cow | Pig | Hen[1] |
| Free | | | | |
| Croneton Sulfoxide | 22 | 0 | 0 | 0 |
| Croneton Sulfone | 2 | 0 | 0 | 0 |
| Croneton Phenol | 0 | 0 | 26 | 3 |
| Phenol Sulfoxide | 3 | 9 | 8 | 23 |
| Phenol Sulfone | 3 | 2 | 6 | 34 |
| Water Solubles | 60 | 87 | 50 | 31 |
| Total | 90 | 98 | 90 | 91 |

[1] Total excreta collected.

TABLE 7

*[$^{14}$C] Croneton equivalents in milk and eggs*

| Metabolites | % of [$^{14}$C] in sample | |
|---|---|---|
| | Milk[1] | Eggs[2] |
| Free | 67 | 88 |
| Croneton Sulfoxide | 39 | 4 |
| Croneton Sulfone | 19 | 4 |
| Phenols | 9 | 79 |
| Unknowns | 0 | 1 |
| Water Solubles | 30 | 5 |
| Unextractable | 3 | 7 |

[1] Milk sample collected 6 hr after single oral dose, 0.5 mg/kg, of [$^{14}$C] ring Croneton.

[2] Eggs sampled on 7th day of treatment of hens; 0.5 mg/kg twice daily.

Because of the very efficient excretion of Croneton by the animals, residues in the tissues were either absent or too low in quantity for characterization. Examination of the milk and eggs (from hens treated daily for 7 days) showed that these products contained [$^{14}$C] residues quite different in nature (Table 7). While a majority of the residues in each substrate was of an organosoluble nature, these free metabolites in milk were comprised largely of Croneton sulfoxide and Croneton sulfone, but were almost entirely Croneton phenol in eggs. Levels of bound, or unextractable, residues were exceedingly small in both milk and eggs.

One could continue to dissect the data generated in the animal studies with Croneton, and continue to reveal interesting similarities and differences among species. However, the most germane points have been emphasized and further comparisons can best be made individually by those with specific points of interest. Generally, it may be concluded that the metabolism of Croneton in animals is quite typical of that expected of an $N$-methylcarbamate containing an ethylthiomethyl substituent on a phenyl ring. While different animal species exhibited metabolic differences which could be toxicologically significant, the rapid rate of excretion of free as well as conjugated Croneton metabolites lessened their potential significance considerably. Thus, the production of free metabolites of sufficient polarity to effect urinary elimination without further metabolism (for example, Croneton sulfoxide in rats) is one characteristic which distinguishes Croneton from most other carbamate insecticides.

ACKNOWLEDGEMENTS

This study was supported by a grant-in-aid from Chemagro Agricultural Division of Mobay Chemical Corporation and, in part, from Environmental Protection Agency Grant No. R802005 and Regional Research Project S-73.

REFERENCES

Kuhr, R. J., Dorough, H. W., "Carbamate Insecticides: Chemistry, Biochemistry and Toxicology", CRC Press, Cleveland, Ohio, 1976.

Nye, D. E., Hurst, H. E., Dorough, H. W., *J. Agric. Food Chem.* 24, 371 (1976).

THE IDENTIFICATION OF p-NITROANILINE AS A
METABOLITE OF THE RODENTICIDE VACOR IN
HUMAN LIVER

J. G. Osteryoung, J. W. Whittaker, J. Tessari and V. Boyes

Department of Microbiology
Institute of Rural Environmental Health
Colorado State University

*ABSTRACT. The liver of a human accidental poisoning victim who injested the rodenticide Vacor was examined for residues of the active ingredient 1-(3-pyridylmethyl)-3-(4-nitrophenyl) urea (RH787). A trace analytical method for RH787 using the polarographic reduction of the nitro function was developed using differential pulse polarography (DPP) which permitted detection of $2 \times 10^{-8}$M RH787 (17 ng RH787/ 3 ml). The nitroaromatic metabolite identified in the sample was p-nitroaniline (PNA), which has polarographic behavior similar to that of RH787. Peak potentials in DPP were: unknown, -571 vM vs SCE; RH787, -585 mV; PNA, -570 mV). Thin layer chromatographic $R_f$ values for the metabolite corresponded to those for PNA. The identification was confirmed by gas-liquid partition chromatography on 5% OV-210 and 1.5% OV-17/1.5% OV-210 columns. The presence of PNA at 5.1 $\pm$ 1.0 mg PNA/kg liver suggests amide hydrolysis of RH787.*

INTRODUCTION

Vacor is a new single dose quick kill rodenticide introduced in the United States market in the summer of 1975 by Rohm and Haas. Results from Vacor trials show it to be effective against rodent species in 4 to 6 hours after a single feeding of as little as 10 mg, in contrast with coumarin rodenticides which require multiple feedings over 4 or more days (Anon., 1975a). Furthermore, Vacor is effective against rats which have developed resistance to coumarin rodenticides. The active ingredient in the Vacor formulation is RH787 (1-(3-pyridylmethyl)-3-(4-nitrophenyl) urea), which is thought to be a nicotinamide antagonist.

$$\text{pyridyl-}CH_2-NH-\overset{O}{\underset{\|}{C}}-NH-\text{phenyl-}NO_2$$

Rohm and Haas have stressed both the target specific rodenticidal properties of Vacor and the relative safety of the active ingredient for other animals and humans (Anon., 1975a). The basis of target specificity and safety claims is that the relatively high $LD_{50}$'s for birds (>500 mg/kg), 2 types of dogs (>500 mg/kg), and Rhesus monkeys (2,000-4,000 mg/kg) are more than ten times as great as the low acute oral $LD_{50}$ reported for several species of rats and mice (Anon., 1974). The $LD_{50}$ reported for the Rhesus monkey tests suggests general insensitivity of primates, including humans, to the poison. Based on these data Vacor will probably be used widely in both urban and rural rodent control programs.

The safety attributes of this rodenticide have been questioned recently because of a series of poisoning incidents in Korea which resulted in seven fatalities (including three children) which were attributed to the ingestion of Vacor as a rice bait (Anon., 1975b). In a related incident, a chemist ingested RH787, had a toxic reaction, but recovered. The question of toxicity of RH787 to children is especially important because of epidemiological evidence which suggests that in the United States almost all coumarin poisonings occur in the 0-4 age group and that these account for about one-third of all pesticide poisonings in that age group (Savage, 1975).

In the case of Vacor, the animal toxicity studies very strongly suggest a low toxicity of the active ingredient to humans, yet the circumstantial evidence of the Korean incidents no less firmly suggests humans are much more sensitive to this substance than are the other primates studied. We have investigated the liver and kidney from an exhumed poisoning victim for traces of RH787. This required the development of an analytical method for RH787.

## ANALYSIS OF HUMAN LIVER SAMPLES

*Analysis by Thin Layer Chromatography*

First we employed a semiquantitative thin layer chromatography (TLC) procedure (Sheasley, 1976). Five gram tissue samples were homogenized with 25 ml of water and 2-12 ml aliquots of the liquid were taken. Each aliquot was salted out with up to 4 g NaCl and extracted with 2 ml tetrahydrofuran (THF). After centrifugation 5 or 10 µl of the THF layer was spotted on silica gel thin layer plates and eluted with 65/35 acetone/benzene. In some cases samples were homogenized with methanol, salted out, centrifuged, and an aliquot of the methanol extract removed and evaporated to about one-fifth the initial volume before spotting.

Development of the TLC plates was done in an iodine chamber and the presence of RH787 confirmed by Ehrlich's reagent. Spiked samples were prepared by grinding dry RH787 standard into the sample prior to homogenizing. Under these conditions, RH787 standards give an $R_f$ value of 0.24 ± 0.03, a brown spot with $I_2$ development, and a lemon-yellow spot with Ehrlich's reagent. THF or methanol extracts of bovine liver samples gave no spots with similar $R_f$ values, while spiked bovine livers and kidneys gave spots with $R_f$ values 0.04-0.08 larger than those of concurrently run standards. THF extracts of liver and kidney samples from the poisoning victim reported to have ingested RH787 gave tenuous spots with $R_f$ values 0.08 larger than those of concurrently run RH787 standards; methanol extracts gave no similar spots but methanol extracts of spiked samples gave spots with $R_f$ values in close agreement with those of concurrently run standards.

The absence of the characteristic lemon-yellow color for RH787 when treated with Ehrich's reagent and the inconclusive detection with iodine indicated the absence of RH787 in the human samples at the sensitivity of the method.

*Spot Test for RH787*

A sensitive spot test for many aromatic nitro compounds, including p-nitrophenyl urea, uses the colored complex formed when 0.1 ml of 10% aqueous tetraethylammonium hydroxide is added to a solution of the sample in 10 ml of dry dimethylformamide (DMF) (Kruse, 1970). When no nitro compound was present, the solution turned a light yellow on addition of the base, while with either RH787 standard or human liver sample extract it turned reddish brown. This result suggested that RH787 or a metabolite containing the aromaetic nitro group is present in the human samples, that the TLC procedure is inadequate with respect to sensitivity, and that other techniques should be tried which would have greater sensitivity and good qualitative identification capability.

*Development of a Differential Pulse Polarographic Method for RH787*

Differential pulse polarography (DPP) is sensitive to $10^{-8} M$ and selective to a particular analyte. This technique measures the concentration of an analyte by measuring the current that flows at a dropping mercury electrode (DME) as a function of the potential imposed on that electrode (Osteryoung and Hasebe, 1976). Frequently this method is insensitive to impurities (including suspended particulate matter) from the substrate.

The analysis for RH787 required a sensitive method which would be unaffected by the lipids, amines and other substances carried into the analyte by the extraction procedure. Furthermore, the structure of RH787 suggested the compound could be polarographically reduced because of the aromatic nitro group. We have developed a DPP method with detection limit $2 \times 10^{-8} M$ (17 ng in 3 ml); the details are published elsewhere (Whittaker and Osteryoung, 1976).

Polarographic characteristics of RH787 are summarized in Table 1. The number of electrons is estimated by comparing the reduction with that of $p$-nitroaniline (PNA), which undergoes a 6 e$^-$ reduction (Fry, 1972).

TABLE 1

*Polarographic characteristics of RH787 in 0.1F TBA(HSO$_4$), pH 2.0 with H$_2$SO$_4$.*

---

$E_{1/2} = -535 \pm$ mV vs SCE

$n = 4 \quad i = 0.66 t^{0.17}$

$D_{DC} = 2.9 \times 10^{-6}$ cm$^2$/sec

$D_{NPP} = 3.1 \times 10^{-6}$ cm$^2$/sec

---

The exponential time dependence of the current, the so-called i-t behavior, can be compared with the 0.17 exponent of time which is predicted by theory for a diffusion controlled process. This indicates that the limiting current is diffusion controlled, and therefore relatively insensitive to matrix effects. The diffusion coefficients (D) can be used to calculate the sensitivity ($\mu A/mM$) for DC or NPP polarography using the Ilkovic or Cottrell equations respectively (Meites, 1965).

Fig. 1 shows differential pulse polarograms of RH787 at high sensitivity. The sensitivity calculated from these data is 236 ± 36 $\mu A/mM$. The 15% relative standard deviation is primarily due to difficulty in estimating the background current at this sensitivity.

## Analysis of Human Liver Extracts

Analysis of multiple extracts of spiked fresh human liver using this procedure showed that the single extraction method described above is not quantitative (Table 2). The procedure outlined in Table 3 was designed for the quantitative recovery of RH787 by extracting the sample 3 times with THF. Because injested RH787 could be transported inside the cells, the cell walls were ruptured by equilibrating the cells under high pressure nitrogen, then rapidly expelling them from a bomb into atmospheric pressure.

TABLE 2

Recovery of RH787 spikes by multiple extraction with THF

|  | 3mg Spike | | 1.5mg Spike | |
| --- | --- | --- | --- | --- |
|  | Recovered RH787, mg | % Found | Recovered RH787, mg | % Found |
| Extract 1 | 2.0 | 67 | 1.1 | 73 |
| Extract 2 | 1.0 | 33 | 0.5 | 35 |
| Total | 3.0 | 99 | 1.6 | 108 |

Using the multiple extraction procedure, recoveries of 107 ± 23% were obtained from samples spiked with RH787 as shown in Table 4. The value of $E_{1/2}$ (-566 ± 11 mV vs SCE) calculated from $E_p = E_{1/2} - \Delta E/2$ was slightly more cathodic

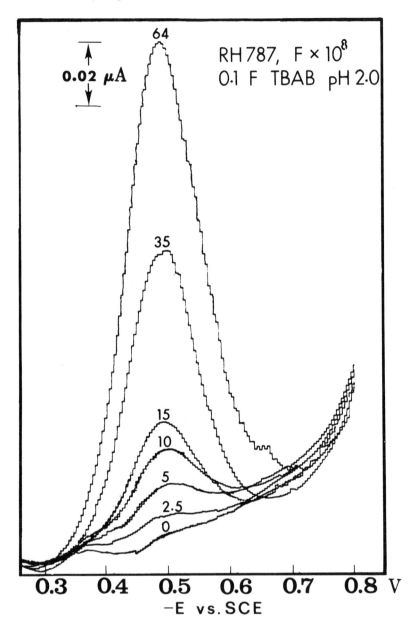

Fig. 1. The differential pulse polarograms of RH787 in 0.1 M tetrabutylammonium bromide at pH 2.0; (0, 2.5, 5, 10, 15, 36, 64) x $10^{-8}$ F RH787; v:2mV/s; $\Delta E$:-100 mV; $t_d$:2S.

than the $E_{1/2}$ observed for the standards. The slight shift in calculated half-wave potential values is not surprising considering the varying amounts of surfactant impurities from the sample.

TABLE 3

Summary of extraction procedure

| | |
|---|---|
| Chop | 2.5-10g liver |
| Blend | with 50 ml $H_2O$ + 20g NaCl |
| Rupture cells | by expelling from PAR bomb at ca 1400 lb/in$^2$ |
| Centrifuge | 20 min at 5000 RFS |
| Remove solids | by filtration |
| Extract | 3 times with 15-20 ml THF |
| Dry | combined extracts with $Na_2SO_4$ |
| Evaporate | THF with $N_2$ |
| Redissolve | in supporting electrolyte |

TABLE 4

Recoveries of RH787 spikes using the multiple extraction method

| SAMPLE | $-E_p$, mV | Added RH787, μg | Recovered RH787, μg (%) |
|---|---|---|---|
| Human Liver | 575 | 2.7 | 2.9 (107) |
| Human Kidney | 550 | 13.6 | 15.0 (110) |
| Human Liver | 542 | 27.0 | 20.7 (77) |
| tlc Extract | 545 | 136.0 | 181.0 (133) |

No peaks were found in this potential region for unspiked bovine liver or for human liver or kidney samples of individuals not exposed to RH787, or from extracts of control areas of TLC plates. Difficulties in background current subtraction account for scatter in the recovery data and contribute to some extent to an apparent shift in the potential of the wave. However, these problems were minor and the method appeared suitable for RH787 analysis.

Human liver extracts were analyzed and found to contain a peak at 571 $\pm$ 6 mv vs SCE which is characteristic of RH787. The DP polarograms in Fig. 2 show a peak at -565 mV which was enhanced upon addition of RH787. The irregularities in the shape of the spiked polarogram were not observed in fresh human or bovine liver extracts and are probably due to the natural surfactants present in the purified liver extracts. TLC clean-up removes these substances as shown in Fig. 3. Based on 8 analyses, 15.2 $\pm$ 3.0 (20%) mg RH787/kg liver was found.

*Polarographic Analysis of TLC Bands*

The polarographic analysis indicated that the liver extract spotted on the chromatographic plate contained about 7 µg RH787. The TLC method was sensitive to about 5 µg RH787 or less. Thus the RH787 spot should have been visible for the chromatographed liver extract. To reconcile this anomaly and to establish quantitative identification of RH787, extracts of the TLC plates were analyzed polarographically.

A preparative TLC plate was spotted with the human liver extract, developed, and the spots detected with fluorescent indicators under UV light. Bands containing the major spots were scraped and extracted with THF. The results of the polarographic analysis of the bands is shown in Table 5. The polarographically active substance migrated farther than the RH787 standard, but had an $R_f$ similar to that of PNA ($R_f$ 0.75). Furthermore PNA, a likely metabolite or degradation product due to amide hydrolysis, had a half-wave potential which was indistinguishable from that of RH787 under the conditions of the analysis. The DP polarograms of the extract of the TLC plate $R_f$ 0.7-0.8 is shown in Fig. 3. This combination of facts suggested that the analyte was PNA.

Various polarographic conditions were checked in hopes of finding conditions which would distinguish between PNA and RH787. Some results are shown in Table 6. The similarity between the reduction potentials of the two compounds under a variety of conditions indicates that the reduction mechanism is such that reduction of the aromatic nitro moiety is largely unaffected by substitution on the para-amino group. The inability to distinguish between the two compounds required another technique to confirm the presence of PHA.

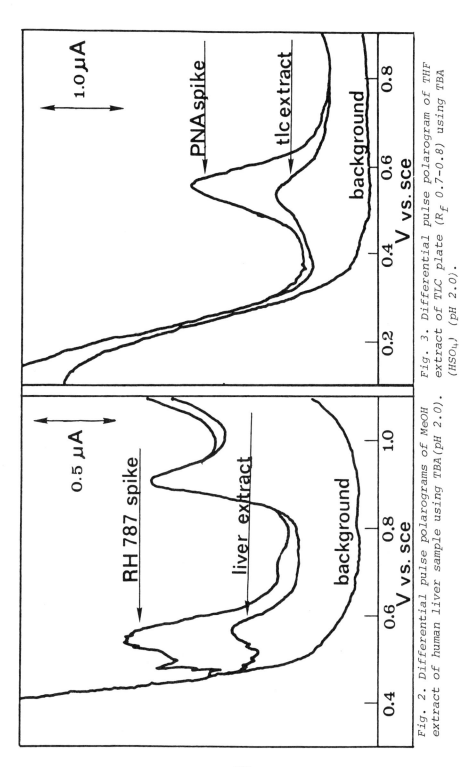

Fig. 2. Differential pulse polarograms of MeOH extract of human liver sample using TBA (pH 2.0).

Fig. 3. Differential pulse polarogram of THF extract of TLC plate ($R_f$ 0.7-0.8) using TBA ($HSO_4$) (pH 2.0).

TABLE 5

Polarographic analysis of extracts of TLC plates

| BAND | $R_f$ range | RH787 $-E_p$, mV | $i_p$, nA | LIVER EXTRACT $-E_p$, mV | $i_p$, nA |
|---|---|---|---|---|---|
| I | 0.15-0.30 | 483 | 50 | - | - |
| II | 0.30-0.55 | - | - | 480(s) | 8 |
| III | 0.55-0.77 | - | - | 480 | 25 |
| IV | 0.77-1.00 | - | - | - | - |

| Location of major spots | $R_f$ | |
|---|---|---|
| Standard RH787 | 0.24 | |
| Liver Extract | 0.45 | 0.73 |

s = shoulder

TABLE 6

Comparison of the reduction potentials for p-nitroaniline and RH787

| MEDIUM | pH | $-E_{1/2}$, mV PNA | RH787 |
|---|---|---|---|
| 0.1F HOAc/NaOAc | 4.6 | 405 | 370 |
| $HSO_4^-$/$Na_2SO_4$ | 3.0 | 310m | 240m |
| 0.1F TBA($HSO_4$) | 2.0 | 520 | 490 |
| 0.25M $H_2SO_4$ | 1.2 | 180m | 112m |
| $H_2SO_4$ | 0.5 | 115m | 78 |

m = maximum

*Gas Chromatographic Analysis for p-Nitroaniline*

RH787 produces broad ill-defined gas chromatographic peaks which are unsuitable for analytical purposes. However, PNA gives sharp, symmetrical peaks under several conditions. The gas chromatograms of an extract of the TLC band ($R_f$ 0.7-0.8) containing the polarographically active substance are

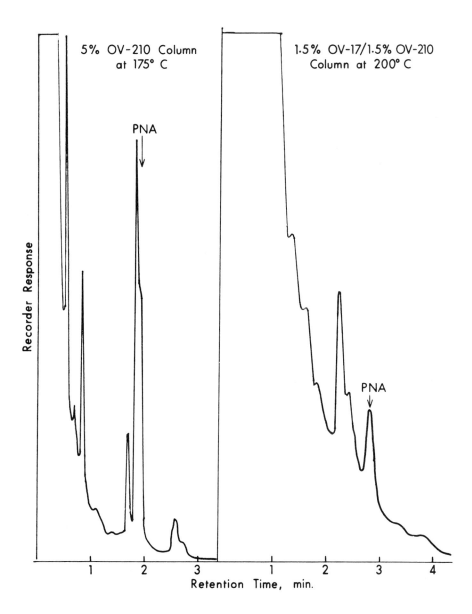

Fig. 4. Gas chromatograms of extract of polarographically active thin layer chromatographic band (Fig. 3). The arrows mark the retention time of p-nitroaniline standards.

shown in Fig. 4. Using the 5% OV-210 column a shoulder was observed at the same retention time as PNA standards (indicated in Fig. 4 with an arrow). When the mixed phase column was used, a clear peak was observed at the PNA retention time.

CONCLUSIONS

From four different tests: 1) two column gas chromatographic analysis, 2) similar TLC retention time, 3) positive spot test for the aromatic nitro group and 4) agreement of polarographic reduction potential with standards and standard additions, it was concluded that PNA was present in the human liver sample.

The amount of PNA present in the liver is $5.1 \pm 1.0$ mg PMA/kg liver based on the previous analysis of extracts of human liver expressed in terms of RH787. The PNA wave is 1.5 times greater than RH787 because 6 electrons are transferred in the former and 4 electrons in the latter. When the quantity is expressed in mg, RH787 (mol wt 272) gives about twice the weight of PNA (mol wt 138) for the same number of moles. Thus RH787 gives 2.96 times the PNA current response.

The amount of PNA found in the liver can be compared to an intravenous mammalian $LD_{50}$ of 40 mg/kg and an oral $LD_{50}$ of 75 mg/kg for birds (Christensen et al., 1975). This indicates PNA is more toxic to mammals or birds than RH787 is. Interestingly enough, the oral $LD_{50}$'s for PNA in rats and mice are reported to be 3,250 and 812 mg/kg respectively (Christensen et al., 1975), which indicates that a metabolic amide hydrolysis to PNA is not the mode of action of the rodenticide. Such speculation only suggests that further investigation of the metabolism of RH787 in rats and mice is indicated since PNA could have been formed during enterrment of the body, possibly by microbial action.

The DP polarographic technique not only has adequate sensitivity and specificity for this problem, but in conjunction with chromatographic separation can provide some evidence regarding the metabolism of RH787. The DPP technique responds to a particular functional group of the molecule. This allows the portion of that molecule containing the functional group intact to be traced during its metabolic breakdown.

ACKNOWLEDGEMENT

This work was supported through a contract with the Epidemiologic Studies Program, Human Effects Monitoring Branch, Technical Services Division, Office of Pesticide

Programs, U.S. Environmental Progection Agency, Washington, D.C. 20460. The mention of trade names or commercial products does not constitute endorsement or recommendation for use by the U.S. Environmental Protection Agency. The views expressed herein are those of the investigators and do not necessarily reflect the official viewpoint of the U.S. Environmental Protection Agency.

REFERENCES

Anon, Technical Bulletin ID-A-156, "Experimental Rodenticide RH787," Rohm and Haas, Philadelphia, Pa. (1974).
Anon, Chem. Eng. News, 24 March, p. 8 (1975a).
Anon, Chem. Eng. News, 7 July, p. 20; 14 July, p. 8 (1975b).
Christensen, H., Luginbyhl, T., Carrol, B. "Registry of Toxic Effects of Chemical Substances," U.S. Department of Health, Education and Welfare's National Institute for Occupational Safety and Health, U.S. Government Printing Office, Washington, D.C. (1975).
Fry, A. J. "Synthetic Organic Electrochemistry," Harper and Roe, New York, N.Y. (1972).
Kruse, J. M. in "The Analytical Chemistry of Nitrogen and its Compounds," C. A. Strevli and P. R. Averell, Eds., Wiley-Interscience, New York. N.Y., p. 452 (1970).
Meites, L., "Polarographic Techniques, 2nd ed," John Wiley and Sons, New York, N.Y., pp. 111-125 (1965).
Osteryoung, J., Hasebe, K., R. Polarog. (Japan) 21, 1 (1976).
Savage, E. P., A Study of Hospitalized Acute Pesticide Poisonings in the United States, 1971-1973. EPA Report. Contracts 68-02-1271 and 68-01-3138. October (1975).
Sheasley, W. D. Rohm and Haas Research Laboratories, Box 219, Bristol, Pa. 19007. Personal Communication (1976).
Whittaker, J., Osteryoung, J., Anal. Chem. 48, 1418 (1976).

# Index

## A

Acceptable daily intake, 9
Anesthesis, for surgical modification, 28
Antipyrine, plasma half-lives in monkeys, 97
Autoradiography, in metabolism studies, 32

## B

Benomyl, fate in animals, 130
$\beta$-BHC, residues in herring gull, 185
Bile
  collection, 26
  diflubenzuron metabolites in, 120
Biopsy techniques, 27
Blood collection
  catheterization, 24
  fetal, 25
  heart puncture, 25

## C

Carboxin, fate in animals, 144
cis-Chlordane, residues in herring gull, 185
Colostomy, of birds, 23
Croneton
  comparative metabolism in animals, 250
  excretion by cow and pig, 239
  metabolites in chickens, 245
  metabolites in cow and pig, 241
  metabolites in milk, 243

## D

DDA, as DDT metabolite
  in deer, 210
  in pig, 180
DDD (TDE)
  as DDT metabolite in deer, 210
  residues in herring gull, 185
DDE
  as DDT metabolite in deer, 210
  as DDT metabolite in pig, 180
  kinetics of elimination from cattle, 165
  residues in herring gull, 185
  urinary metabolite of
    in pig, 179
    mass spectrum of, 179
DDT
  fate in deer, 193
  in feces of deer, 207
  metabolism in Rhesus monkey, 93
  residues in herring gull, 185
  residues in tissues of deer, 200, 206
  urinary metabolites of, in pig, 180
DDT and metabolites, kinetics of elimination from cattle, 167

Deer, mortality from DDT treatment, 212
N-(3',5'-Dichlorophenyl)-succinimide, fate in animals, 143
2,4-Dichlorphenoxy acetic acid
  fate in humans, 63
  metabolism in animals, 54
  residues in meat, 66
  residues in milk, 65
Dieldrin
  kinetics of elimination from cattle, 167
  residues in herring gull, 185
Diamidafos
  excretion by rats, 222
  IR spectra of, and metabolites, 226
  mass spectra of, and metabolites, 224
  metabolites in cattle, 229
  metabolites in plants, soil, water, 219
  pharmacokinetics in rats, 220
Diflubenzuron
  fate in stable fly and house fly, 122
  mass spectra, of metabolites, 121
  metabolism, in cattle and sheep, 116
Dimethirimol, fate in animals, 139
Dosing techniques
  dermal, 21
  inhalation, 20
  intratracheal, 21
  intravenous, 20
  oral
    gelatin capsule, 19
    ingesta—exchange, 18
    oesophageal cannulae, 19
    simulated-meal, 19
    stomach tube, 19
    rumen puncture, 19
Drazoxolon, fate in animals, 141

## F

Facilities, for metabolism studies, 18
Feces collection, 22
Fistulation, of rumen, 28
Fuberidazol, fate in animals, 133

## G

Griseofulvin, fate in animals, 150

## H

Hexachlorobenzene
  kinetics of elimination from cattle, 167
  residues in herring gull, 184
Heptachlor expoxide
  mass spectrum of, 186
  residues in herring gull, 185
Heptachlorobiphenyl methyl ether, mass spectrum of, 190
Herring gull, organochlorine residues in, 183
Hexachlorobiphenyl methyl ether, mass spectrum of, 190
Hexachlorophene, fate in animals, 148

## I

Intestinal loops, use in metabolism studies, 29

## J

Juvenile hormone and analogs, fate in animals, 112

## L

Liver homogenate, in metabolism studies, 32

## M

MCPA
  metabolism in animals, 54
  residues in meat, 66
  residues in milk, 65
Metabolism stalls, 21
Methoprene, fate in animals, 114
Methoxychlor, residues in herring gull, 185
Mirex, residues in herring gull, 184

## N

*p*-Nitroaniline, as vacor metabolite in human liver, 260
No-effect level, of pesticides, 9
*trans*-Nonachlor
  mass spectrum of, 186
  residues in herring gull, 185

## O

Oesophageol fistulation, of ruminants, 30
Organ perfusion, in metabolism studies, 31
Oxychlordane
  mass spectrum of, 187
  residues in herring gull, 185

## P

PBB, kinetics of elimination from cattle, 167
PCB
  kinetics of elimination from cattle, 167
  residues in herring gull, 184
Pentachlorobiphenyl methyl ether, mass spectrum of, 189
Pentachloronitrobenzene, fate in animals, 146
Pesticides, comparative toxicity to rodents, 129
Photo-*cis*-chlordane, mass spectrum of, 187
Photo-*cis*-nonachlor, residues in herring gull, 185
Photomirex, residues in herring gull, 184
Phenoxy herbicides
  comparative metabolism of, 53
  metabolic pathways, 58
  pharmacodynamics, 59
  toxicity to animals, 57
Primates
  as research animals, 79
  classification, 81
  comparative metabolism with other mammals, 86
  foreign compound metabolism by liver preparations *in vivo*, 87
  similarities among species, 84
  insecticides and enzyme induction, 89

## R

R-20458, fate in animals, 113
Regulatory aspects of pesticide metabolism, 47
Residues
  crop tolerances, 8
  quantitation of, 7

Respiratory gases, collection of, 23
Reticulum, cannulation of, 28
Rhesus monkeys
 DDT effects on liver oxidases, 92
 liver biopsy, 99
 monooxygenase activities in liver, 100
Rumen
 artificial rumen techniques, 30
 cannulation, 28
 isolated pouch, 29

## S

Silvex
 fate in humans, 63
 metabolism in animals, 54
 residues in meat, 68
 residues in milk, 65

## T

Tetrachlorobiphenyl methyl ether, mass spectrum of, 189

Theoretical daily intake, 9
Thiabendazole, fate in animals, 135
Thiophanate–methyl, fate in animals, 138
2,4,5-Trichlorophenoxy acetic acid
 fate in dogs, rats, and humans, 62
 metabolism by animals, 54
 residues in meat, 68
 residues in milk, 65
Toxicity
 evaluation of oral, 4
 of plant metabolites, 5
"Toxicological significance," of pesticide metabolism, 5

## U

Urine collection, 22